Selections from
The Distribution and
Abundance of Animals

Looking back, I think it was more difficult to see what the problems were than to solve them.

DARWIN

Selections from
The Distribution and
Abundance of Animals

With a new Preface

H. G. Andrewartha
and
L. C. Birch

The University of Chicago Press
Chicago and London

The University of Chicago Press, Chicago 60637
The University of Chicago Press, Ltd., London

89 88 87 86 85 84 83 82   1 2 3 4 5

Library of Congress Cataloging in Publication Data

Andrewartha, H. G. (Herbert George), 1907–
   Selections from The distribution and abundance of
animals.

   Bibliography: p. 255
   Includes index.
   1. Animal ecology.   2. Animal populations.   I. Birch,
L. Charles, 1918–.   II. Title.
QH541.A522   1982   591.5   82–6948
ISBN 0–226–02031–2   AACR2
ISBN 0–226–02032–0 (pbk.)

Contents

Preface

In solving ecological problems we are concerned with what
animals do *in their capacity as whole, living animals, not as dead
animals or as a series of parts of animals. We have next to study the
circumstances under which they do these things, and, most important of all, the limiting factors which prevent them from doing certain other things. By solving these questions it is possible to discover
the reasons for the* distribution and number of animals in nature.

ELTON (1927, p. 34)

POPULATION ecology is the study of the distribution and abundance of
animals in nature. This perspective was first clearly set out by Elton
(1927) and was elaborated in *The Distribution and Abundance of Animals*
(1954). Despite the passage of nearly three decades since the publication
of this book, it is still in demand. It seems reasonable to suppose that this
demand might be more satisfactorily met by reprinting some selections which
exclude those parts of the original that seem to have aged most. We hope
that by publishing this selection of "readings" we might preserve and make
more accessible those parts of the original that seem to us to be still relevant
to modern theory and knowledge of population ecology. Much water has
passed under the ecological bridge since 1954, and our own ideas have developed along new lines, particularly with respect to the theory of environment and what we now call "the multipartite population." So we have written a new book which presents these new ideas. It is called *The Ecological
Web;* it will be available shortly after this book of readings has been
published.

The two books are complementary because *Selections* essentially covers
those topics from *The Distribution and Abundance of Animals* that have not
been repeated in detail in the new book because they have not advanced much
during the intervening years (chapters 3, 6, and 13); *Selections* also presents
the beginnings of the modern theory about environment and the multipartite
population (chapters 12 and 14).

In *Selections,* chapter 3 on the innate capacity for increase in numbers still serves as an introduction for many students to the concepts of "animal demography." Ultimately the distribution and abundance of animals depends upon the chance the animal has to survive and to reproduce in the particular places in which it lives. The innate capacity for increase is a statistic that tells a great deal about the chance to survive and to reproduce. This was the first chapter to be written on the subject in an ecological text apart from A. J. Lotka's *Elements of Physical Biology* (1925). But Lotka wrote primarily about man, for at the time he wrote, demographic concepts had hardly as yet been applied to animal ecology in general. The concepts of our chapter still stand. What has happened since 1954 has been an extension of the application of these concepts to many more species of animals and to plants as well (e.g., Harper, 1977) and the use of computers to do the calculations.

In our theory of environment (1954), we had four components of environment: weather, food, other animals and organisms causing disease, and a place in which to live. Temperature, which is a typical component of weather (others are moisture and light), is the subject of the second chapter of this book (chapter 6). The way of studying it has not changed substantially over the years. This chapter tells how the ecologist studies its influence both in the laboratory and in nature.

Although "a place in which to live" as a component of environment has been subsumed in other components of environment (Andrewartha, 1970), chapter 12 is important because it shows the extreme finicalness of animals with respect to the sorts of places where they live. This contributes largely to the patchiness of local populations, which is a concept developed further in the last chapter.

"Empirical Examples of the Numbers of Animals in Natural Populations" (chapter 13) comprises twelve case histories. These are examples of how ecology is studied. They still stand as examples of sound practical ecology.

The last chapter, "A General Theory of the Number of Animals in Natural Populations," was the first comprehensive attempt to develop a theory of animal numbers in natural populations that was based on the concept of an environment that included everything that influenced the animal's chance to survive and reproduce. Up to that time, most theorists had been protagonists of one part or another of what we saw as the whole environment.

Chapters 3 and 6 are included because modern advances in these fields have not seriously outdated them. Chapters 12 and 14 are included because they reflect the condition thirty years ago of certain important modern ideas, the concept of multipartite natural population and the theory of spreading the risk. Chapter 13 earns a place because these ecological case histories retain their intrinsic merit irrespective of whether they are interpreted within the framework of an old-fashioned or a modern theory of environment.

The chapters in *Selections* retain the numbers they had in *The Distribution and Abundance of Animals,* so that instead of chapters 1 through 5 we have, as noted above, chapters 3, 6, 12, 13, and 14. We hope this effort to reduce the cost of these reprinted chapters will not distract the reader. The text of the chapters is unchanged from the original chapters. So it contains some references to sections and figures in chapters that are not reproduced in this book. They can of course be found in the original book. Pages have been renumbered throughout, and the new indexes and bibliography refer to the new pages in the *Selections*.

CHAPTER 3

The Innate Capacity for Increase in Numbers

In looking at nature, it is most necessary to keep the foregoing considerations always in mind, never to forget that every single organic being may be said to be striving to the utmost to increase in number, that each lives by a struggle at some period of its life, that heavy destruction inevitably falls either on the young or the old, during each generation or at recurrent intervals. Lighten any check, mitigate the destruction ever so little and the number of the species will almost simultaneously increase to any amount.

DARWIN, *The Origin of Species*

3.0 INTRODUCTION

ANY animal living in a particular environment may be expected to grow at a certain rate, to live for a certain period, and to produce a certain number of offspring, usually spread over a certain span of its life. For any one species, each individual in the population will have its own particular speed of development, longevity, and fecundity at different ages in its life. It is usually more useful, however, to speak of the mean values for the population. There will be a mean rate of growth of individuals in the population; a mean longevity, which is more usefully considered as a distribution of ages at which different individuals die; and a mean fecundity, which is more usefully considered as mean birth-rates at different ages of the mothers. The values of these means are determined in part by the environment and in part by a certain innate quality of the animal itself. This quality of an animal we may call its *innate capacity for increase*. We shall devote a full chapter to the discussion of this concept, because it is important in relation to the distribution and abundance of animals in nature and because, despite its importance, it has hitherto received but little attention. Animal ecologists, in stressing the environment, have tended to overlook the innate qualities of the animal or else, recognizing the need to consider the animal, have insisted on a sharp line of division between the "biotic potential" of the animal and the "resistance" of the environment. This was a mistake, for the animal's innate capacity for increase cannot be considered apart from its environment.

1

Nevertheless, the innate capacity for increase is a quality which is just as characteristic of the species as is, for example, its size. It is, however, a character that is more difficult to measure and define, since it may vary widely in different environments. The analogy with size may be carried a little further, for it is well known that for some animals the size may vary with such components of the environment as temperature, moisture, food, and so on. Nevertheless, as a rule, ordinary variations in the environment may not make much difference to the size of the animal, and it is often sufficient to define size without any special reference to the environment from which the animal has come. On the other hand, relatively small changes in one or another component of the environment may result in enormous differences in the animal's innate capacity to increase; so when this character is being considered, it is always necessary to define very carefully the particular environment in which the animal is living.

Environments in nature rarely, if ever, remain constantly favorable or constantly unfavorable but fluctuate irregularly between the two extremes. The animal's innate capacity for increase fluctuates correspondingly, being sometimes positive and sometimes negative. While conditions remain favorable and the innate capacity for increase remains positive, the numbers increase. If it remained so indefinitely, the species would continue to multiply until eventually it covered the earth. The elephant probably has the lowest innate capacity for increase of any animal. Even so, calculations based on the assumption of continuous increase lead to the conclusion that the progeny of one pair of elephants could populate the earth within several hundred years. Darwin estimated that a single pair of elephants could give rise to 19,000,000 elephants after a period of 750 years. But elephants do not increase like this. Adverse weather, food shortages, and a limitation of the number of suitable places in which elephants can live impose checks to increase; from time to time the rate of increase becomes zero or negative. Once the checks are lifted, the population will increase again. The rate of multiplication will be determined by the innate capacity for increase.

The concept of environment which we use in this book is described in chapter 2. Environment is considered to apply to individuals in a population but not to a population as a whole. The latter usage, though common, is so limited as to lead to endless confusion. The environment of individuals in a population is partly the other animals in the population. There is no clear-cut division between the population and its environment. But the distinction between the *individual* and its environment is clear-cut and can be defined with precision. This idea is very important in relation to the present chapter. It is desirable for the sake of simplicity to speak of the birth-rate or the death-rate within a population. This need not be confusing if it is remembered that the death-rate of a population may be found by integrating the expectation of life of all the individuals that comprise it. It is also made clear in chapter 2 that any environment may be analyzed into four general sets of components, namely, weather, food, other animals, and a place in which to live. Each one of these will be discussed in detail in Part III. In nature, one or several components may predominate to determine the *actual* rate of increase, which we shall call *r*. But in an experiment it is possible to exclude predators, diseases, and all other organisms of different kinds; and food, space, and other animals of the same kind can be artificially kept at optimal levels. Temperature, moisture, and the other components of weather and the quality of food cannot be excluded from the experiment in the same way, but, in an experiment, they can be kept artificially constant. We define r_m, the innate capacity for increase, as the maximal rate of increase attained at any particular combination of temperature, moisture, quality of food, and so on, when the quantity of food, space, and other animals of the same kind are kept at an optimum and other organisms of different kinds are excluded from the experiment. This is an approximate definition, because it does not mention the distribution of ages in the population. The precise definition is given in section 3.132.

3.1   THE INNATE CAPACITY FOR INCREASE

Biologists have sometimes referred to "the reproductive rate" as the characteristic defining the capacity of a species to increase in numbers. But, without the additional information of the death-rate (or survival-rate) of the offspring, this tells little about the innate capacity of the species to increase. Darwin pointed this out in *The Origin of Species*. There is no relation between the numbers of eggs laid by animals and the abundance of these animals in nature. The number in nature depends upon a balance struck between birthrate and death-rate. It has been a characteristic feature of quite a number of ecological studies, particularly those with insects, to study one or another characteristic of the species, such as fecundity, speed of development, longevity, and so on. From such information, inferences have been made about distribution and abundance. It is quite possible that a study of one particular function alone may lead to different conclusions from those which might be drawn from the study of some other function (e.g., Holdaway, 1932). It is necessary rather to consider the sum-total effect of all those functions which make for increase in numbers and which make one species differ from another, that is to say, all those functions which influence birth-rate and survival-rate.

An animal's innate capacity for increase depends upon its fecundity, longevity, and speed of development. With a population, these are measured by the birth-rate and the survival-rate (or its inverse, the death-rate). When the birth-rate exceeds the death-rate, the population increases in numbers at a rate dependent upon the difference between them. When they are the same, the population numbers remain stable. When death-rate exceeds birth-rate, the population declines at a rate dependent upon the difference between them. This argument is simple enough. But complexity is introduced as soon as we seek to estimate quantitatively the rate at which the population increases or decreases. *The difficulties are encountered because both the number of births and the probability of death vary with the age of the animal.*

The students of human populations were the first to appreciate this principle, and Lotka (1925) derived a function which he called "the intrinsic rate of natural increase" to take into account the changes in birth-rates and death-rates with age. Demographers have coined certain expressions which we shall also use. The table which sets out the detailed distribution of birth-rate with age is called the "age-schedule of births," and the detailed distribution of deaths with age is called the "age-schedule of deaths." The particular birth-rate and death-rate which are characteristic of a particular age-group are called the "age-specific birth-rate" and the "age-specific death-rate." The "intrinsic rate of natural increase" is sometimes called the "Malthusian parameter," because of the emphasis given by Malthus to the geometric rate with which populations could theoretically increase in numbers. We shall not use the expression "Malthusian parameter" because it is misleading. There is a precise meaning in the science of biometry for the word "parameter." The symbol r_m is quite clearly a "statistic" and not a "parameter" (see R. A. Fisher, 1946, *Statistical Methods for Research Workers*, p. 7). If we knew the exact value of the "intrinsic rate of natural increase" of an animal, we would call that a "parameter" and specify it by a Greek letter. We cannot, in fact, know the parameter exactly, but we can make an estimate of its value. This estimate is termed a "statistic." There is a sense in which every reproductive pair of individuals in the population has its "intrinsic rate of natural increase," but we are always interested in the estimate of a mean value for the population. This is the statistic r_m, which we call the "innate capacity for increase."

Although r_m was originally devised for the study of human populations (Lotka, 1925), its relevance to ecology was at least implicit in A. J. Lotka's book, *Elements of Physical Biology*, published in 1925, and in R. A. Fisher's *The Genetical Theory of Natural Selection*, published in 1930. To Leslie and Ranson, of the Bureau of Animal Population in Oxford, however, must be attributed the credit of having adapted the statistic r_m for a more general use in animal ecology in their experimental determination of r_m for the vole *Microtus agrestis* (Leslie and Ranson, 1940). Since then its use has been extended to insect populations by Birch (1948, 1953*a*, *b*), Leslie and Park (1949), Evans and Smith (1952), and Howe (1953*a*, *b*). And we may expect that it will come to be more widely used by ecologists.

Before considering the way in which the appropriate tables of birth-rates and death-rates may be constructed, it is best to consider the relationship between birth-rates, death-rates, and the innate capacity for increase, r_m.

3.11 *The Relationship between Birth-Rate, Death-Rate, and the Innate Capacity for Increase*

The rate of increase of a population which has an assumed constant age-schedule of births and deaths and which is increasing in numbers in an un-limited space is given by.

$$\frac{\delta N}{\delta t} = bN - dN$$
$$= (b - d)N,$$

where t denotes time and b and d are constants representing the instantaneous birth-rate and death-rate. Now $b - d$ is the infinitesimal rate of increase which is the innate capacity for increase, r_m. Hence

$$\frac{\delta N}{\delta t} = r_m N.$$

This is the differential form of the equation describing the curve of geometric increase of an infinitely expanding population.

In the integrated form:

$$N_t = N_0 e^{r_m t}$$

where

$N_0 = $ number of animals in time zero,
$N_t = $ number of animals in time t,
$r_m = $ innate capacity for increase,
$e = $ base of Naperian logs.

Since we are dealing with geometric increase, it follows that the relationship between $\log_e N$ and t is linear. The slope of this line is the value r_m. This is made clear by the following:

$$N_t = N_0 e^{r_m t};$$

thence

$$\log_e N_t = \log_e N_0 + r_m t \quad \text{(since Naperian log of } e = 1)$$
$$= a + r_m t \quad \text{(where } a \text{ is a constant).}$$

This is the equation for the straight line with co-ordinates $\log_e N$ and t. The infinitesimal rate of increase r_m should not be confused with the finite rate of increase, i.e., the number of individuals added to the population per head per week. An example will make the distinction clear. Consider a population which multiplies ten times in every 2 weeks. The infinitesimal rate of increase may be shown to be 1.1513 by the following calculations:

Since $N_t = N_0 e^{r_m t}$,   then $\dfrac{N_t}{N_0} = e^{r_m t}$,

and $\therefore r_m = \dfrac{\log_e(N_t/N_0)}{t} = \dfrac{\log_e 10}{2} = 1.1513$ per head per week.

The finite rate of increase of the same population is 3.16 per head per week, as shown by the following. Let the finite number of individuals arising from one female in one week be λ. Then, by definition,

$$\lambda = \frac{N_t}{N_0} \quad \text{when} \quad t = 1;$$

$$\text{but} \quad \frac{N_t}{N_0} = e^{r_m} \quad \text{(when } t = 1\text{)},$$

$$\text{i.e.,} \quad \lambda = e^{r_m}$$
$$= \text{antilog}_e \; r_m.$$

In this example $\lambda = \text{antilog}_e \; 1.1513 = 3.16$ per head per week. Hence a population which multiplies ten times in every 2 weeks has an infinitesimal rate of increase of 1.1513 and a finite rate of 3.16 $(= \sqrt{10})$. That is to say, the population multiplies 3.16 times per female per week. It follows that, by the end of the second week, one individual will have given rise to $(3.16)^2 = 10$ individuals, and by the end of the third to $(3.16)^3$, and so on.

3.12   *The Construction of Age-Schedules of Births and Deaths*

The generalized relationship between birth-rate, death-rate, and r_m was discussed in section 3.11 without any consideration of the complete expression of both birth-rates and death-rates in terms of age-groups in the population. The actual calculation of r_m, however, involves a knowledge of these details.

The birth-rate of a population is best expressed as an "age-schedule of births." This is a table which gives the number of offspring (or eggs in the case of insects) produced in unit time by a female aged x. This is usually designated m_x, where the suffix x denotes the age-group.[1] In these calculations we usually count only the females. If equal numbers of males and females are born, then m_x is the total number of eggs or offspring for each age-interval, divided by 2. The nature of this statistic will be clear from an examination of Tables 3.01 and 3.02 and Figures 3.01 and 3.02, which show, among other things, the age-schedule of births for two very different sorts of populations, a vole and an insect (Leslie and Ranson, 1940; Birch, 1948).

The m_x values for the vole in Table 3.01 give the mean number of live daughters born per 8 weeks per female of ages shown in the x column. The female vole begins to breed when it is about 3 weeks old. In Table 3.02 the m_x values for the weevil *Calandra oryzae* are given in terms of the total eggs laid per female per week, divided by 2, since the sex-ratio is unity. In each case

---

[1] A glossary of terms used in this chapter is given at the end of the chapter.

the values of x refer to the mid-point of each age-group. The intervals for the age-groups may be chosen quite arbitrarily and depend partly on the method by which the data were obtained. Birth is taken as zero age in the case of the vole, and the time the egg was laid is taken as zero age in the case of the

TABLE 3.01*

LIFE-TABLE, AGE-SPECIFIC FECUNDITY-RATES, AND NET REPRODUC-
TION-RATE (R_0) OF THE VOLE *Microtus agrestis* WHEN REARED IN
THE LABORATORY

Pivotal Age in Weeks (x)	l_x	m_x	$l_x m_x$
8	0.83349	0.6504	0.54210
16	.73132	2.3939	1.75071
24	.58809	2.9727	1.74821
32	.43343	2.4662	1.06892
40	.29277	1.7043	0.49897
48	.18126	1.0815	0.19603
56	.10285	0.6683	0.06873
64	.05348	0.4286	0.02292
72	0.02549	0.3000	0.00765

$$R_0 = 5.90424$$

* After Leslie and Ranson (1940).

TABLE 3.02*

LIFE-TABLE, AGE-SPECIFIC FECUNDITY-RATES, AND NET REPRODUCTION-RATE (R_0)
OF RICE WEEVIL *Calandra oryzae* AT 29° C. IN WHEAT OF 14 PER CENT MOISTURE
CONTENT

Pivotal Age in Weeks (x)	l_x	m_x	$l_x m_x$	
0.5				
1.5				⎫ Immature stages:
2.5	0.90	....	....	⎬ egg, larva, and
3.5				⎭ pupa
4.5.......	.87	20.0	17.400 ⎱	
5.5........	.83	23.0	19.090	
6.5......	.81	15.0	12.150	
7.5......	.80	12.5	10.000	
8.5......	.79	12.5	9.875	
9.5......	.77	14.0	10.780	
10.5......	.74	12.5	9.250	
11.5......	.66	14.5	9.570 ⎬ Adults	
12.5......	.59	11.0	6.490	
13.5......	.52	9.5	4.940	
14.5......	.45	2.5	1.125	
15.5......	.36	2.5	0.900	
16.5.......	.29	2.5	0.725	
17.5.......	.25	4.0	1.000	
18.5......	0.19	1.0	0.190 ⎰	

$$R_0 = 113.485$$

* After Birch (1948).

insect. The fecundity-table for *C. orzyae* shown in Table 3.02 and Figure 3.02 is quite typical of many insects. In contrast to the vole, the sexually immature stages of the insect (egg, larva, pupa, and early adult) occupy a greater proportion of the total life-span. In both the case of mammals such as the vole and insects such as the weevil, the data can be obtained experimentally by observing a number (or "cohort") of individuals from the time they are born until they die and recording the number of offspring they produce as they grow older. Similar information for human populations cannot be obtained in this manner, as individuals do not always remain in the same place for such long-term observations. The age-schedule of fecundity of a

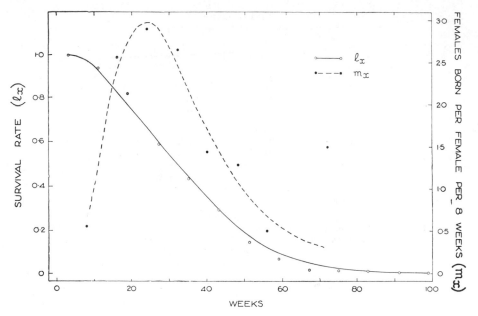

Fɪɢ. 3.01.—The life-table and age-specific fecundity curves for the vole *Microtus agrestis* reared in the laboratory. (After Leslie and Ranson, 1940.)

human population can, however, be compiled from census figures, which give details of the age at which women in the community give birth to offspring (Pearl, 1924).

The offspring from the parent-generation die at different ages. The table which gives this information is known as the "life-table." Again we usually count only the females; but life-tables can, of course, be compiled for males as well. The life-table gives the probability at birth of being alive at age x. This probability is usually designated l_x. At zero age ($x = 0$) it is designated l_0 and is taken as having a value of unity ($l_0 = 1$). The second columns of Tables 3.01 and 3.02 and Figures 3.01 and 3.02 show the life-tables for the vole *Microtus agrestis* and the weevil *C. oryzae*. These life-tables are not complete but cover the reproductive period of life, which is all that is necessary to consider in studying rates of increase. Referring to Table 3.02, the figures $x = 4.5$ and $l_x = 0.87$ mean that, from a sample of 100 eggs of zero age, 87 per cent survive to the mid-point between the fourth and fifth week. Between the eighteenth and nineteenth weeks, only 19 per cent are still alive. In the case of both the vole and the weevil the life-table was obtained by following through the survival of a sample of individuals from birth until the last member of the population died.

Life-tables were first used by students of human populations. It is only in recent years that ecologists have begun to realize their importance. There are,

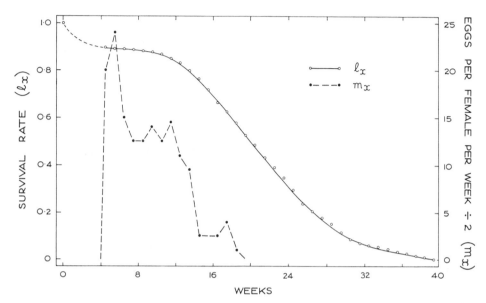

FIG. 3.02.—The life-table and age-specific fecundity curves for the weevil *Calandra oryzae* at 29° C. in wheat of 14 per cent moisture content. The curves provide all the information necessary for the estimation of the innate capacity for increase. (After Birch, 1953*a*.)

consequently, not many complete life-tables for other species. The information which is available has been largely determined since 1935, when Pearl and Miner (1935) compared the life-tables of the few organisms which had then been studied with any degree of completeness (Deevey, 1947). From the stand-point of this chapter we are primarily interested in life-tables of animals reared artificially in the laboratory. The life-table of populations in nature may be quite different, for in nature the population is exposed to numerous hazards which will destroy members of the population in different age-groups, depending upon what are the chief causes of deaths. These life-tables have been called "ecological life-tables." Such information is generally much more difficult to determine than experimental life-tables. Deevey (1947) has adequately re-viewed the available information on ecological life-tables.

It is sometimes convenient to express the life-table in forms other than a survivorship or l_x table. Although alternative modes of expressing the same information do not concern us in the calculation of the innate capacity for increase, brief mention will be made of two alternatives of particular use in certain population studies. These are the d_x and the L_x curves. The curve giving number of deaths within age-intervals is know as the d_x curve. The L_x curve, also known as the life-table age-distribution or life-table age-structure, gives the proportion of individuals alive which are between ages x and $(x + 1)$ (Dublin, Lotka, and Spiegelman, 1949):

$$L_x = \int_x^{x+1} l_x \delta x.$$

In practice it is usually adequate to ignore the curvature between age-groups; hence

$$L_x = \frac{l_x + l_{x+1}}{2}.$$

The life-table age-distribution assumes special importance in the consideration of age-structure of populations, as will be shown later in this chapter.

3.13   *Calculation of the Innate Capacity for Increase*

The innate capacity for increase was first worked out in the study of human populations. Because of the relatively large part of the human life-span occupied by pre-reproductive development and the relatively low birth-rate, it is sufficient to use an approximate arithmetical procedure in calculating r_m. With small mammals and insects, in which the birth-rate and length of life are likely to be quite different from those for human populations, a more precise solution is needed. We shall follow the historical approach and give the approximate solution first.

3.131   APPROXIMATE CALCULATION OF r_m

We have seen how the geometric increase of a population is given by the equation

$$N_t = N_0 e^{r_m t},$$

where N_0 is the number of reproducing individuals at time t_0, N_t is the number of reproducing individuals at time t, and r_m is the innate capacity for increase. This relationship provides us with a means of calculating the innate capacity for increase from the age-schedule of births and the age-schedule of survival-rates (life-table), i.e., the m_x and l_x columns of Tables 3.01 and 3.02. In calculating r_m from these data, it is convenient to deal in time-intervals of generations. Now the number of individuals at the end of a generation will be

$$N_T = N_0 e^{r_m T},$$

where T is the mean length of a generation. Hence

$$\frac{N_T}{N_0} = e^{r_m T}.$$

Now N_T/N_0 is simply, by definition, the ratio of total female births in two successive generations. It is the multiplication per generation or *net reproduction-rate*. It is usually designated R_0. Thus

$$R_0 = e^{r_m T}$$

and

$$r_m = \frac{\log_e R_0}{T}.$$

The intrinsic rate of natural increase can thus be determined if the net reproduction-rate and the mean length of a generation are known. Both these statistics can be readily estimated from the tables giving the age-schedule of births and survival-rates. The calculations involve nothing more complicated than simple arithmetic.

The method of estimating R_0 is set out in Tables 3.01 and 3.02. The $l_x m_x$ products for each age-group are summed, the total being the value of R_0, i.e., $R_0 = \Sigma l_x m_x$. The net reproduction-rate of the vole, for example, is 5.904. In other words, a population of voles has the capacity to multiply 5.904 times in each generation.

The estimation of T is also straightforward. The mean duration of a generation is defined as the mean period elapsing from birth of parents to birth of offspring. Clearly, this is only an approximate definition, since offspring are not born at any one time but over a period, e.g., in the vole from about the fourth to the seventy-second week of life. Both the mean age of the mother at which offspring are born and the extremes of age at which they are born vary for each individual in the population. But we may consider the births for each generation as concentrated at one moment, with successive generations spaced T units apart, where T is the mean duration of a generation (Dublin and Lotka, 1925). It may be defined approximately as follows:

$$T = \frac{\Sigma l_x m_x x}{\Sigma l_x m_x}.$$

The figures for the product $l_x m_x$ in the last column of Tables 3.01 and 3.02 may be regarded as a frequency-distribution. The mean of the distribution is the approximate value of T. In these particular examples:

$$M.\ agrestis:\quad T = \frac{143.75}{5.904} = 24.4 \text{ weeks,}$$

$$C.\ oryzae:\quad T = \frac{941.85}{113.49} = 8.3 \text{ weeks.}$$

Knowing R_0 and T, we can determine the innate capacity for increase. In the example of the weevil *C. oryzae* (Table 3.02) the estimate of r_m is made as follows:

$$R_0 = 113.49, \quad T = 8.30 \text{ weeks},$$

$$r_m = \frac{\log_e 113.49}{8.30} = 0.56.$$

Now this estimate of r_m is approximate only, since the procedure outlined for estimating the mean length of a generation (T) was approximate only. The true estimate of T necessitates a preliminary calculation of r_m, as can be seen from the following relationship:

$$R_0 = e^{r_m T}.$$

Hence

$$T = \frac{\log_e R_0}{r_m}.$$

In some cases the first approximation to r_m obtained by the above method may be accurate enough. It is sufficiently accurate for most estimates of rates of increase of human populations. With animals which have a greater rate of increase, this method may be too crude. This is the case in the example of the weevil *C. oryzae*. We described the direct method first, because it illustrates quite simply the relationship between birth-rates, survival-rates, mean length of a generation, net reproduction-rate, and the innate capacity for increase.

A glance at Figures 3.01 and 3.02 will show the importance of the nature of the curves for birth-rate and survivorship in determining the innate capacity for increase of a population. The maximal rate of increase would be given when the peaks of both curves occur at the same point on the same time-axis. If the birth-rate curve were moved farther to the right along the time-scale, the capacity of the population to increase in numbers would be lower. It is, in fact, possible to calculate exactly the outcome of moving one curve along the time-axis with respect to the other and so have a precise measure of the way in which the two components, birth-rate and survival-rate, determine the innate capacity of a population to increase in numbers.

3.132   PRECISE CALCULATION OF r_m

Lotka (1925) showed mathematically that the distribution of ages in a population in which the birth-rates and death-rates for each age-group remain constant and which is increasing in unlimited space would approach a certain distribution which he called "the stable age-distribution" because it would not vary with time. Lotka also showed that, as such a population approached its "stable age-distribution," its rate of increase also approached a certain constant, which he called the "the intrinsic rate of increase." This is the quantity which we have called the "innate capacity for increase," r_m. It will now be seen that the definition of r_m given in section 3.0 was approximate, for it did not specifiy that the population should have a stable age-distribution. The precise value of r_m may be obtained by solving the following equation:

$$\int_0^\infty e^{-r_m x} l_x m_x \, \delta x = 1,$$

where 0 to ∞ is the life-span of the reproductive stages and l_x and m_x are as defined on pages 36, 38. The accurate solution of this equation involves some tedious calculation. Various approximations are usually permissible which reduce the calculations to simple arithmetical procedures. The product $l_x m_x$ is known for each age-group (see Tables 3.01 and 3.02). Various values of r_m can be substituted in the equation, and the true value obtained by graphical interpolation (Leslie and Ranson, 1940; Birch, 1948). When the calculation was made for the weevil *C. oryzae*, the value of r_m was found to be 0.76, to be compared with 0.56 obtained by the approximate method. Further simplifications in the procedure of estimating r_m for insects have been introduced by Howe (1953a); these modifications enable estimates to be made with greater speed and with an accuracy which is sufficient for most purposes. The usual method of calculating r_m for human populations may be found in Dublin and Lotka (1925, appendix) or Lotka (1939, p. 68).

3.2   THE STABLE AGE-DISTRIBUTION

By definition, the innate capacity for increase, r_m, is the actual rate of increase of a population with stable age-distribution. The actual rate of increase depends upon the distribution of ages in the population. Since the distribution of ages changes with time in all populations other than those in which the distribution is the stable age-distribution, the actual rate of increase will also change with time. This will be true even though the values for l_x and m_x (the birth-rates and death-rates in the different age-groups) remain constant. For these reasons, r_m is the only statistic which adequately summarizes the physiological qualities of an animal which are related to its capacity for increasing. This is the reason for keeping the word "innate" in the definition of r_m. For the same reasons, r_m is the only statistic which serves to compare different species with respect to these qualities. The actual rates of increase which ignore the distributions of ages in the populations are no good at all for these purposes.

The stable or Malthusian age-distribution may be calculated from the life-table and the innate capacity for increase. Thus if C_x is the proportion of the population of stable age-distribution aged between x and $x + \delta x$ and b is the instantaneous birth-rate,

$$C_x = b e^{-r_m x} l_x.$$

The instantaneous birth-rate can be calculated from the following equation:

$$\frac{1}{b} = \int_0^\infty e^{-r_m x} l_x \, \delta x.$$

For the usual methods of computation, reference should be made to Dublin and Lotka (1925) and Leslie and Ranson (1940). Leslie has, however, pointed out another method of calculation which saves much of the numerical integration involved in the more usual methods (Birch, 1948). If at time t we consider a stable population consisting of N_t individuals and if during the interval of time t to $t + 1$ there are B_t female births, we may define a birth-rate as follows:

$$\beta = \frac{B_t}{N_t}.$$

Then, if we define for the given life-table (l_x) the series of values L_x by the relationship

$$L_x = \int_x^{x+1} l_x \, \delta_x$$

(the stationary or life-table age-distribution of the actuary), the proportion (P_x) of individuals aged between x and $x + 1$ in the stable population is given by

$$P_x = \beta L_x e^{-r_m(x+1)},$$

$$\frac{1}{\beta} = \sum_{x=0}^m L_x e^{-r_m(x+1)},$$

where $x = m$ to $m + 1$ is the last age-group considered in the complete life-table age-distribution. It will be noticed that the life-table (l_x) values for the complete age-span of the species are required for the computation of P_x and β. But where r_m is high, it will be found that, for the older age-groups, the terms $L_x e^{-r_m(x+1)}$ are so small and contribute so little to the value of β that they may be neglected.

The calculations involved are quite simple and are illustrated in the following example for *C. oryzae* at 29° C. (Table 3.03). In the present example, instead of calculating the values of L_x, the values of l_x were taken at the mid-points of each age-group. This was considered sufficiently accurate in the present instance. It should also be pointed out that, whereas only the total deaths among immature stages were required in the calculation of r_m, the age-specific mortality of the immature stages is needed for the calculation of the stable age-distribution. In this example 10 per cent of the immature stages died—and 98 per cent of these deaths occurred during the first week of larval life (Birch, 1945a). Hence the approximate value of L_x for the mid-point of the first week will be 0.95 and thereafter 0.90 for successive weeks of the larval and pupal period (second column, Table 3.03). The stable age-distribution is shown in the fifth column of Table 3.03. This column simply expresses the fourth column of figures as percentages. It is of particular interest to note the high proportion of immature stages (95.5 per cent) in this theoretical population. This is as-sociated with the high value of the innate capacity for increase. It emphasizes a point of practical importance in estimating the abundance of insects such as *C. oryzae* and other pests of stored products. The number of adults found in a sample of wheat may be quite a misleading representation of the true size of the whole insect population. Methods of sampling are required which will take account of the immature stages hidden inside the grains, such, for example, as the "carbon dioxide index" developed by Howe and Oxley (1944). The

TABLE 3.03*

CALCULATION OF THE STABLE AGE-DISTRIBUTION OF *Calandra oryzae* AT 29° C. WHEN $r_m = 0.76$

Age-Group (x)	L_x	$e^{-r_m(x+1)}$	$L_x e^{-r_m(x+1)}$	Percentage Distribution $100\beta L_x e^{-r_m(x+1)}$	
0............	0.95	0.4677	0.4443150	54.740	95.5 per cent
1............	.90	.2187	.1968300	24.249	total imma-
2............	.90	.10228	.0920520	11.341	ture stages
3............	.90	.04783	.0430470	5.304	
4............	.87	.02237	.0194619	2.398	
5............	.83	.01046	.0086818	1.070	
6............	.81	.00489	.0039609	0.488	
7............	.80	.002243	.0017944	0.221	
8............	.79	.001070	.0008453	0.104	4.5 per cent
9............	.77	.000500	.0003850	0.047	total
10............	.74	.000239	.0001769	0.022	adults
11............	.66	.000110	.0000726	0.009	
12............	.59	.000051	.0000301	0.004	
13............	.52	.000024	.0000125	0.002	
14............	0.45	0.000011	0.0000050	0.001	
		$1/\beta = 0.8116704$		100.000	

* After Birch (1948).

nature of this stable age-distribution has a bearing on another practical problem. It adds further evidence to that developed from a practical approach (Birch, 1946a), as to how it is possible for *C. oryzae* to cause heating in vast bulks of wheat, when only a small density of adult insects is observed. It is a reasonable supposition that the initial rate of increase of insects in bulks of wheat may approach the innate capacity for increase and therefore that the age-distribution may approach the stable form. Little, however, is known about the actual age-distribution in nature at this stage of an infestation. Howe (1953a) made an estimate of the age-distribution of *C. oryzae* in a bin of wheat at the peak of an infestation and found that 11 per cent of the population were adults. This is fairly close to the theoretical estimate of 4.5 per cent of adults in a stable age-distribution at 29° C., which is shown in Table 3.03.

The stable age-distribution of the vole is shown for comparison, together with its life-table or stationary age-distribution in Table 3.04. Conditions under which the life-table distribution might be expected will be discussed in chapter 9. It will be noticed that the older age-groups are not represented in Tables 3.03 and 3.04. This does not mean that there would be none of these old animals living in those populations but that the dilution of the population with young is so great that in a random sample of a thousand or so the probability of including a vole older than 56 weeks or a weevil older than 9 weeks is small. Tables 3.03 and 3.04 indicate the sort of age-distribution which is approached by a rapidly growing population when the density of the population in terms of numbers per unit of space is relatively low.

TABLE 3.04*
STABLE AGE-DISTRIBUTION AND LIFE-TABLE AGE-DISTRIBU-
TION OF THE VOLE *Microtus agrestis*.

Age-Group	Stable Age-Distribution	Life-Table Age-Distribution
0.................	57.7	23.5
8.................	25.5	21.2
16.................	10.7	17.8
24.................	4.1	13.8
32.................	1.4	9.7
40.................	0.5	6.3
48.................	0.1	3.8
56.................	....	2.1
64.................	....	1.0
72.................	....	0.5
80.................	....	0.2
88.................	....	0.1
	100.0	100.0

* After Leslie and Ranson (1940).

3.3   THE INSTANTANEOUS BIRTH-RATE AND DEATH-RATE

We have seen that the rate of increase of a population at any time may be given by the difference between its crude birth-rate and its crude death-rate. We have also shown that this difference does not give the innate capacity of the population to increase in numbers, because the age-distribution of the population has not been taken into account. But the capacity of the population to increase in numbers is adequately measured by the rate of increase that the population would have if it had a stable age-distribution. This also gives an adequate basis of comparison for different populations. A population with a stable age-distribution has, of course, a birth-rate, b, and a death-rate, d. It is the difference between these two quantities which we have designated the innate capacity for increase, r_m:

$$r_m = b - d.$$

In the process of estimating the innate capacity for increase, neither the birth-rate nor the death-rate is directly estimated. They may, however, be estimated. once r_m is known.

We have already defined a birth-rate by the expression

$$\frac{1}{\beta} = \sum_{x=0}^{m} L_x e^{-r_m(x+1)}.$$

This is not, however, the same as the instantaneous birth-rate, b. The relationship between the two is given by

$$b = \frac{r_m \beta}{e^{r_m} - 1}.$$

Thus, in the example for *C. oryzae*, we have $1/\beta = 0.81167$ (Table 3.03), $r_m = 0.76$, and thus $b = 0.82$; and the difference between r_m and b is the instantaneous death-rate, $d = 0.06$.

3.4 THE INFLUENCE OF ENVIRONMENT ON THE INNATE CAPACITY FOR INCREASE

It is now clear why the innate capacity for increase cannot be expressed quantitatively except for a particular environment. In the case of poikilothermic animals such as insects, temperature and moisture are two of the important components which influence longevity and fecundity and therefore the innate capacity for increase. Temperature may be less important for homoiotherms, though for any particular homoiotherm it is hardly possible at this stage of our knowledge to say just how temperature might influence its capacity to increase in numbers. Any component of the environment (other than than those which have been excluded by definition—see sec. 3.0) may influence the value of the innate capacity for increase. We shall give three simple illustrations of this principle.

TABLE 3.05

INFLUENCE OF TEMPERATURE AND MOISTURE ON FINITE RATE OF INCREASE (λ)* OF *Calandra oryzae* (SMALL "STRAIN") AND *Rhizopertha dominica* IN WHEAT AT CERTAIN COMBINATIONS OF TEMPERATURE AND MOISTURE CONTENT

	C. oryzae						*R. dominica*				
TEMP. (°C.)	Per Cent Moisture Content of Wheat					TEMP. (°C.)	Per Cent Moisture Content of Wheat				
	14	12	11	10.5	10		14	11	10	9	8
13.0......	0	0	0	0	0	18.3......	0	0	...	...	0
15.2......	1.007	0	0	0	0	22.0......	...	1.11	0	...	0
18.2......	1.15	...	1.10	0	0	26.0......	...	...	1.28	...	0
23.0......	1.54	1.35	...	...	.	32.3......	1.99	...	...	...	0
25.5......	1.83	...	1.36	1.07	0	34.0......	2.14	...	1.47	1.38	0
29.1......	2.15	1.50	1.42	0.96	0	36.0......	...	...	1.30	0	0
32.3......	1.65	...	...	0	0	38.2......	1.34	1.19	0	0	0
33.5......	1.13	1.00	0	0	0	38.6......	0	0	0	0	0
34.0......	...	0	0	0	0						

* $\lambda = e^{r_m}$ = rate of increase per female per week. After Birch (1953a).

Temperature and moisture both influence the innate capacity for increase of these two grain beetles, *C. oryzae* and *R. dominica*. In many ways these two beetles are suitable for laboratory studies; it is possible to determine their age-specific fecundity and survival-rates at various combinations of temperature and moisture. From this information the innate capacity for increase can be calculated for various combinations of temperature and moisture within the range in which the species survive. There is a zone of temperature and moisture within which the capacity for increase is greatest. There is a zone beyond which no increase can occur at all. This is shown in Table 3.05 and is illustrated in Figure 3.03. The rate of increase has been plotted in the figure as a finite rate of increase (λ). This is the multiplication of the population per female per week.

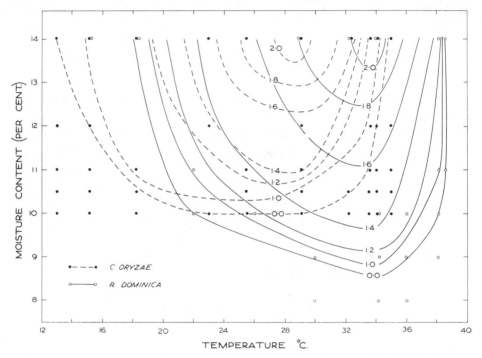

FIG. 3.03.—The finite rate of increase (rate of increase per female per week λ) of *Calandra oryzae* (small "strain") and *Rhizopertha dominica* living in wheat of different moisture contents and at different temperatures. The points on the graph show combinations of temperature and moisture at which experiments were done; the values of λ at these combinations are given in Table 3.05. The lines have been drawn through equal values of λ. (After Birch, 1953a.)

It is a rather more familiar concept than the infinitesimal rate of increase, r_m. These values have been plotted on a graph with temperature and moisture as co-ordinates, and lines have been drawn through points of equal value of λ. The isopleth for unity defines the zone of temperature and moisture in which populations could maintain their numbers but could not increase ($r_m = 0$). When λ is less than unity (r_m is negative), the death-rate exceeds the birth-rate, and the population eventually dies out. The time it takes to become extinct will depend upon the value of λ. The zero isopleth ($r_m = -\infty$) shows the zone in which no insects of reproductive age are added to the population; if any eggs are laid, the death-rate is 100 per cent in the immature stages. For purposes of our present discussion, we can confine our attention to finite rates of increase. For other purposes, it is more convenient to consider the infinitesimal rate of increase, r_m. The relationship between r_m and λ over the range of values of λ shown in Figure 3.03 are summarized in Figure 3.05. This figure permits one expression to be readily changed to the other.

From Figure 3.03 we deduce that *R. dominica* might be expected to thrive better than *C. oryzae* in wheat that is hot and dry. But *C. oryzae* might be able

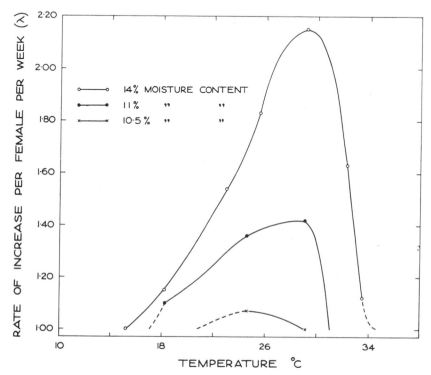

Fig. 3.04.—The finite rate of increase (λ) of *Calandra oryzae* (small "strain") living in wheat of different moisture contents and at different temperatures. (After Birch, 1953a.)

to multiply in wheat so cold that *R. dominica* could not maintain a population there. We might also expect the geographic distributions of the species to be related to climate in a way that is indicated by the position of the isopleths for zero-increase in Figure 3.03. The observed distributions of the two species in Australia agree with this hypothesis: *C. oryzae* is a serious pest wherever wheat is stored; *R. dominica* is also a pest in the warmer parts of the mainland but is virtually absent from Tasmania, which is colder.

If we were to consider only the information given in Figure 3.03, we might expect that populations would be able to persist in any place where the temperature and the moisture content of the wheat exceeded the values indicated by the isopleths for zero on the graph. But we would expect the rate of increase to be greater at certain combinations of temperature and moisture content than at others. The rate of increase is conveniently measured by λ. Values of λ for *C. oryzae* calculated for a number of combinations of temperature and moisture are shown in Figure 3.04. Reference back to Figure 3.03 shows how the two species differ in these respects. For example, at a temperature of 33° C. and in wheat of 12.5 per cent. moisture content, *C. oryzae* would take 16 weeks to reach a population of the size which *R. dominica* could reach in 10 weeks. The range

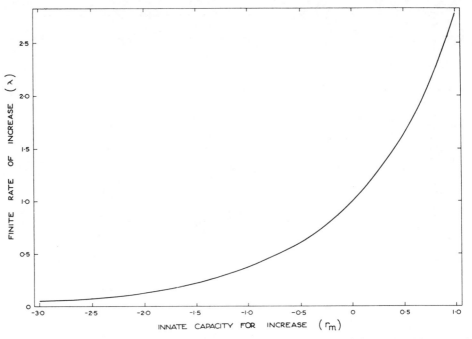

FIG. 3.05.—The relationship between the innate capacity for increase (*r_m*) and the finite rate of increase (λ).

of λ which is positive for the two species extends from λ = 1 to λ = 2.15. For the beetle *Ptinus tectus* the range is less, extending from λ = 1 (at about 13° C.) to a maximum of λ = 1.331 at 27° C. (Howe, 1953*a*). At first glance, neither of these ranges may seem great. However, quite small differences in the value of λ will cause big differences in the numbers which would be reached after several weeks. This can be readily appreciated by looking at Figure 3.06, which shows the multiplication per female after different numbers of weeks for different values of λ.

Another instructive feature of a graph such as Figure 3.03 is that it enables us to see at a glance the combinations of temperature and moisture at which two species have the same capacities for increase. These are shown by the intersections of isopleths of the two species. This information is tabulated in Table 3.06.

When we come to consider natural populations, it is necessary to take into account certain checks to increase which were artificially excluded from the experiments upon which Figure 3.03 was built. With these graminivorous species, predators are unimportant; they are not as a rule, influenced by the

TABLE 3.06
COMBINATIONS OF TEMPERATURE AND MOISTURE AT WHICH
Rhizopertha dominica and *Calandra oryzae* HAVE SAME FINITE
RATE OF INCREASE ($\lambda = e^{r_m}$)

λ	Temp. (°C.)	Moisture Content of Wheat (Per Cent)
1.0	23.5	10.3
1.2	21.5	12.0
1.4	24.5	11.7
1.6	28.3	12.8
1.8	30.0	13.0

distribution of their stocks of food because they can usually rely upon men to carry them to where their food is. But a colony breeding freely in a bin of wheat may be destroyed when the bin is emptied; or a period may be set to their multiplication by a change in the weather. If catastrophes come with

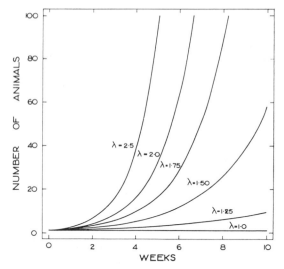

FIG. 3.06.—The relationship between the finite rate of increase (rate of increase per female per week λ) and the size of the population resulting from one female.

equal frequency in two places, then, other things being equal, the animals will be more numerous in the place where their innate capacity for increase is greater. Alternatively, in a place where λ is low, a much longer interval between catastrophes may be required if the animals are to become so numerous that the species may be considered "common" in that area.

Extending this idea to cover comparisons between different species whose requirements and behavior are essentially similar but which differ with respect to their innate capacities for increase, we might say that the common species will be those which have a sufficiently high value of λ to enable them to reach large numbers in the time available between catastrophes. The rare species will be those in which the innate capacity for increase is low relative to the intervals which are allowed them for multiplication between catastrophes. This hypothesis is supported by the values of r_m estimated by Howe (1953a) for 9 species of Ptinidae. One of these beetles, *Ptinus tectus*, is regarded as a serious pest of stored wheat in Britain. Some of the others are widely distributed, but none is numerous enough to be considered a serious pest. A glance at Table 3.07 shows that r_m was greatest for *P. tectus*, which is also the most numerous species in nature. The four species which follow *P. tectus* in Table 3.07 gave

TABLE 3.07
INNATE CAPACITY FOR INCREASE, r_m, AND FINITE RATE OF INCREASE (λ)*
AT 25° C. FOR 9 SPECIES OF PTINIDAE

Species	r_m	λ
Ptinus tectus...................	0.395	1.462
Gibbium psylloides...............	.235	1.265
Trigonogenius globulus............	.227	1.255
Stethomezium squamosum..........	.178	1.195
Mezium affine...................	.160	1.173
P. fur..........................	.094	1.099
Eurostus hilleri..................	.072	1.075
P. sexpunctatus.................	.044	1.045
Niptus hololeucus................	0.043	1.044

* $\lambda = e^{r_m} =$ rate of increase per female per week. After Howe (1953a).

values for λ which were about 18 per cent less than that for *P. tectus*. A value for λ of 1.25 indicates a tenfold increase in just over 10 weeks. But none of these species is considered a serious pest. A colony of one of these species, left undisturbed in a large bin of warm grain, would undoubtedly become numerous in due course. The fact that, in nature, none of these species is ever found in large numbers suggests that a period is set to their increase by the weather, by the emptying of the bin, or by some other "catastrophe" which is likely to be repeated before the colony has had time to become numerous. The risk for *Eurostus hilleri* may be even greater, for its innate capacity for increase is smaller still. During the 13 years which have elapsed since it was first discovered in Britain, it has spread widely over the country, but nowhere has it become abundant.

Birch (1953a) compared the innate capacity for increase of the small and the large "strains" of *C. oryzae* in wheat and in maize. Table 3.08 shows that the small "strain" had a higher value of λ than the large "strain" in wheat, but the reverse was true in maize. In grain storages in Australia the small "strain" is common in wheat and the large "strain" is common in maize. But the small "strain" is so rare in maize and the large "strain" so rare in wheat that they have not yet been recorded as breeding in these in grain stores. Again there is a correlation between the value of λ at a favorable temperature and the abundance of the species.

A parallel might be drawn between the beetles which we have discussed in this section and the grasshoppers which were discussed in section 1.1. We explained how periods of drought or excessive humidity might reduce a large population of grasshoppers to a small remnant, which would multiply again when the weather became more kindly. With the insects which live in stored grain, the catastrophe is more likely to come when the bin is emptied, so that the grain may be used; or during periods of prolonged storage the grain may become intolerably hot (Birch, 1946a). In each case a period is set to increase by recurrent catastrophes which may wipe out the whole population or reduce it to a remnant. This is characteristic of many natural populations (chap. 13).

The concept of innate capacity for increase which we have discussed in this chapter is an abstraction from nature. In nature we do not find populations

TABLE 3.08
FINITE RATE OF INCREASE* FOR THE SMALL AND THE LARGE
"STRAINS" OF *Calandra oryzae* IN WHEAT AND MAIZE AT 29.1° C.
AND 70 PER CENT RELATIVE HUMIDITY

"Strain"	Wheat	Maize
Small	2.15	1.52
Large	1.76	1.55

* Rate of increase per female per week, $\lambda = e^{r_m}$. After Birch (1953a).

with stable age-distributions: weather fluctuates; food may be scarce or sparsely distributed; and an animal's chance to survive and multiply usually depends on how many and what sorts of other animals are associated with it. For these reasons the actual rate of increase r, which may be observed in a natural population and which we discuss at length in subsequent chapters, is much more complex than the theoretical r_m of this chapter. In chapter 9 we show how r departs from r_m under the influence of too few or too many other animals of the same kind. In chapter 10 we discuss other organisms of different kinds, chiefly with respect to their influence on one component of r, namely, the death-rate. The theory of innate capacity for increase is basic to all these discussions, although, of course, the statistic r_m, as we have defined it, does not have any real meaning in relation to a natural population.

Our definition of r_m is arbitrary. We might have defined it broadly so that it was influenced by all the components of environment. This would amount to saying that r_m is the actual rate of increase which would be observed in a natural population if only the ages in the population were distributed as in the stable age-distribution. There would be no profit in such a broad definition because there would be no way in which we could measure r_m experimentally.

Nor could we study it in natural populations, because natural populations probably do not have stable age-distributions.

It is better to limit the meaning of the innate capacity for increase arbitrarily as we have done, because then r_m can be measured precisely by experiment. So long as its limitations are remembered, it may then form a powerful aid to further thought about the distribution and numbers of animals in nature.

GLOSSARY OF TERMS

THE m_x TABLE—This gives the age-schedule of female births (or eggs destined to become females). For any particular parental age-group of pivotal age x, m_x is the number of female births.

THE l_x OR LIFE-TABLE—This gives the age-schedule of survival. For any particular age-group of pivotal age x, l_x is the proportion of individuals alive at the beginning of the age-interval.

THE d_x TABLE—This gives the age-schedule of mortality. It is another way of expressing the life-table. For any particular age-group of pivotal age x, d_x is the proportion of individuals dying within this age-interval.

THE NET REPRODUCTION RATE, R_0—This is the multiplication per generation. It is expressed as the ratio of total female births in two successive generations.

THE MEAN DURATION OF A GENERATION, T—This is the mean time from birth of parents to birth of offspring.

THE INNATE CAPACITY FOR INCREASE, r_m—This is the infinitesimal rate of increase of a population of stable age-distribution (sec. 3.0).

THE FINITE RATE OF NATURAL INCREASE, λ—This is the multiplication per female in unit time of a population of stable age-distribution. This is best defined by the equation $\lambda = e^{r_m}$.

THE STABLE AGE-DISTRIBUTION—This is the age-distribution which would be approached by a population of stable age-schedule of birth-rate and death-rate (i.e., m_x and l_x constant) when growing in unlimited space.

THE L_x TABLE OR LIFE-TABLE AGE-DISTRIBUTION—This gives the proportion of individuals alive between the ages x and $x + 1$ in the life-table. It is the age-distribution of a population which is stationary, the total number of births being constant and equal to the number of deaths over the same period and having an unchanging age-schedule of death-rates.

CHAPTER 6

Weather: Temperature

Climate plays an important part in determining the average numbers of species, and periodical seasons of extreme drought or cold, I believe to be the most effective of all checks.

DARWIN, *The Origin of Species*

6.0 INTRODUCTION

DARWIN considered that in "the struggle for existence" the indirect influence of weather operating through the food supply was more important than its direct influence on the animals themselves. There can be no doubt that weather may, in this indirect way, greatly influence the animal's chance to survive and multiply. We shall have something to say about this elsewhere, particularly in chapters 13 and 14. For the present we are concerned with the direct influence that weather may have on the speed of development, fecundity, and longevity of animals; these are the three components of the innate capacity for increase, r_m (sec. 3.1). Weather may also influence the dispersal of animals and other aspects of their behavior (secs. 5.1 and 6.22). It is convenient to consider, first, the independent influence of the separate components of weather, temperature, humidity, light, and so on and then later to try to describe how, in nature, the influence of one may depend on its interaction with the others. Temperature and humidity have been widely studied. Light, air movements, and atmospheric pressure have also been shown to influence development or behavior in particular cases. The volume of the literature in this field is now enormous. Some of it has been summarized and listed by Allee *et al.* (1949). Our selection of examples is restricted to the few which are sufficient to illustrate our central theme, namely, the principles which govern the distribution and numbers of animals in nature and the kinds of studies which bring these out most clearly.

6.1 THE RANGE OF TEMPERATURE WITHIN WHICH AN ANIMAL MAY THRIVE

It is well known that animals of one sort or another thrive high on mountains or at high latitudes and in the heart of equatorial continental deserts. The extremes of temperature from the hottest to the coldest places where animals live are great. It has been claimed that the larvae of certain Diptera thrive at temperatures as high as 55° C. or even higher; but we shall probably be near the mark if we accept the opinion of Brues (1939) that 52° C. is about the highest temperature at which any animal is known to thrive indefinitely. At the other extreme, there is the beetle *Astagobius angustatus*, which is known to carry on its life-cycle in ice grottos, where the temperature range is between

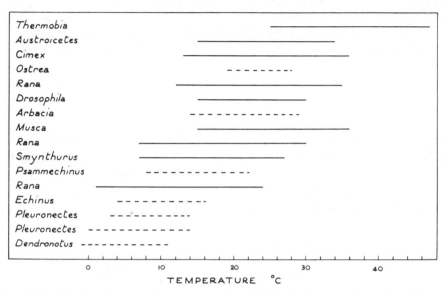

Fig. 6.01.—Showing the range of temperature favorable for development of a variety of aquatic (*broken lines*) and terrestrial or amphibious animals (*solid lines*). Note (*a*) that the ranges for the aquatic animals are mostly much shorter than for the terrestrial ones and (*b*) the wide variation between species in the limits of the favorable range; for example, the upper limits of the favorable range for *Dendronotus*, which lives in the ocean at high latitudes, is below the lower limit for *Thermobia*, which inhabits bakehouses. The three lines for *Rana* refer to three different species. (Data taken largely from Moore, 1940*a*.)

— 1.7° and 1.0° C. (Allee *et al.*, 1949). But no individual species is known which can thrive over such a wide range as from 0° to 50° C. The range of temperature favorable to any particular species is related to the prevailing temperatures in the places where the animal usually lives. Not only the average temperature but also its variability will be reflected in the physiology of the particular animal. Those that live in cold places (or are active during the cold season in warm temperate zones) have a favorable range lower than that for animals from warmer zones. For example, the favorable range for the development of the eggs of *Smynthurus viridis* (Collembola) extends from 7° to 27° C. (Davidson, 1931); but with *Austroicetes cruciata* (Orthoptera) the favorable range for the development of eggs extends from 15° to 34° C. (Birch, 1942). The former develop during the winter in the warm, temperate climate of southern Australia, the latter develop during the spring in a more northerly (somewhat warmer) part of the same general climatic zone. This principle is further illustrated in Figure 6.01, in which the species are arranged in order of the median temperature of their favorable ranges.

It will be noticed in Figure 6.01 that all the animals with narrow ranges (16° C. or less) were aquatic, living in places where the difference between minimal and maximal temperature is small, and that most of those with wide ranges (20° C. or more) were terrestrial or amphibious, living in places where the temperature is more variable. Moore (1940*b*) tabulated the favorable range for 41 aquatic animals from among Crustacea, Echinodermata, Mollusca, Tunicata, and Pisces and found that for 37 of them (91 per cent) the range was 16° C. or less; for 17 (42 per cent) the range was less than 14°; and 2 fishes were restricted to the narrow range of 8°. Six amphibians and 2 terrestrial insects listed by Moore had ranges varying from 22° to 25° C.; and he cited numerous other terrestrial arthropods and amphibians for which the range, though not determined precisely, was known to exceed 20° C.

If an animal is exposed to a low, or a high, temperature which is outside the limits of the favorable range, it may be killed directly, or it may continue to live for an indefinite period, yet fail to grow or produce any young. The result for the population is much the same in either case. These two aspects of "unfavorable" temperatures will be discussed separately: the former in sections 6.3 ff., and the latter in sections 6.2 ff. in relation to the influence of temperature, within the favorable range, on behavior and "activity," speed of development, and fecundity.

6.2   THE INFLUENCE OF TEMPERATURE WITHIN THE FAVORABLE RANGE

6.21   *The Preferred Temperature, or "Temperature-Preferendum"*

When animals are allowed to move freely along a temperature gradient, they usually congregate between quite narrow limits of temperature. This narrow band of temperature has been called the "preferred temperature" or the "temperature-preferendum." For example, when Doudoroff (1938) placed a number of the fish *Gisella nigricans* in a temperature gradient extending from 20° to 30° C., they arranged themselves at the different temperatures with the frequencies shown in Figure 6.02. Most of the fish congregated at a temperature near 26°; ordinates from the abscissa at 25° and 28° inclose about 75 per cent of the total area under the curve, indicating that about three-quarters of the fish came to rest within this 3° band of temperature; over 60 per cent stayed between the limits 25° and 27° C. The preferred temperature varied, up to a certain ceiling, directly with the temperature at which the fish had been living. Beyond this ceiling the preferred temperature became independent of the temperature to which the fish had been acclimatized. This is illustrated by Figure

6.03, which is redrawn after Fry (1947), who called this upper limit above which the preferred temperature cannot be raised by further increasing the temperature at which the fish had been living the *final preferendum*. He considered this to be a useful characteristic to measure for ecological studies.

Natural populations may not be homogeneous with respect to the "preferred temperature." For example, Wilkes (1942) found a trimodal distribution for this character in a population of the chalcid parasite *Microplectron*. The

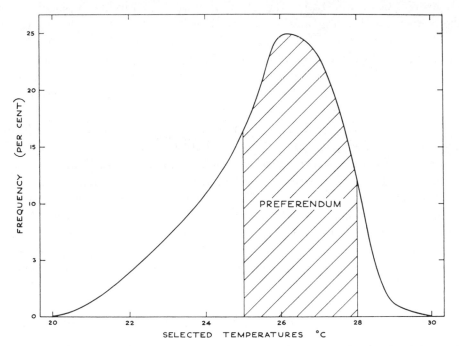

Fig. 6.02.—Showing the "preferendum" for the fish *Gisella nigricans*. The height of the graph indicates the proportion of the fish choosing each temperature. About 75 per cent of the fish congregated in the temperature indicated by the shaded area. (Modified after Doudoroff, 1938.)

greatest frequency was at about 25° C., with two minor ones at about 15° and 8°. By selecting and inbreeding the wasps which congregated at the lower temperature, he produced a distinct race in which the distribution was virtually unimodal: this race had a "preferendum" at about 9° C. (sec. 15.111). These results are decisive because of the large numbers of animals tested and the large differences obtained. Very often experiments with preferred temperatures with terrestrial animals are more difficult to interpret than those done with aquatic animals. The reason for this is that whenever there is a gradient in temperature, there are also, inevitably, gradients in the relative humidity or the saturation deficit (or both) of the atmosphere (sec. 7.234).

Indeed, Wellington (1949*b*), after a thorough investigation of the zones of temperature and humidity preferred by larvae of the moth *Choristoneura* *fumiferana*, came to the conclusion that it was not influenced by temperature to any extent that could be detected. Instead, the situations in which they congregated were determined entirely by the evaporative power of the atmosphere in these zones (sec. 7.12; Fig. 7.02). Wellington then examined the (quite extensive) literature dealing with the experimental aspect of this subject and concluded that, with few exceptions, the results attributed to temperature were obtained without the influence of humidity being adequately eliminated

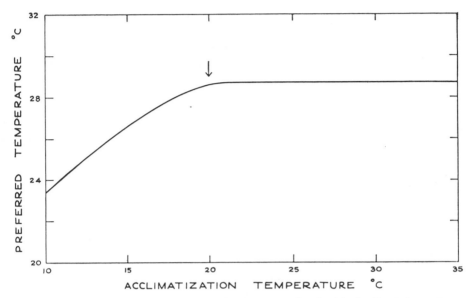

Fig. 6.03.—Showing the "final preferendum" temperature for the fish *Gisella nigricans*. For explanation see text. Note how the "final preferendum" increases with the temperature of acclimatization up to a limit and then becomes independent of the temperature of acclimatization. (Data from Doudoroff, 1938.)

in the experimental design. This may be so, but we find it hard to accept the implication that temperature is usually so unimportant as it seems to be with *C. fumiferana*, however difficult it may be to demonstrate an independent response to temperature in cases where the influence of humidity is dominant. In addition to the more recent papers by Wellington (1949a, b, c), the subject has been reviewed by Deal (1941) and Fraenkel and Gunn (1940).

The question as to whether temperature or humidity exercises the greater independent influence on the animal seeking a "preferred" zone in which to rest is not easily resolved by ordinary field observations on behavior. Because changes in absolute humidity are slow and small relative to changes in temperature, fluctuations in the evaporative power of the atmosphere, whether measured directly, as "evaporation" or indirectly as "saturation deficit" (sec. 7.234), follow fluctuations in temperature closely (Fig. 6.04). When observational data, such, for example, as the behavior of a swarm of locusts, is found to be correlated with temperature, the correlations with evaporation and saturation deficit will be of the same order. This follows from the simple rule in statistics which states that if two (or more) variables x and z are closely correlated positively and if a third variable y is correlated with x, then the correlation between y and z will be of the same order and sign as that between y and x. This does not preclude the possibility of a relationship between y and x which is independent of that between y and z and vice versa; but more compli-

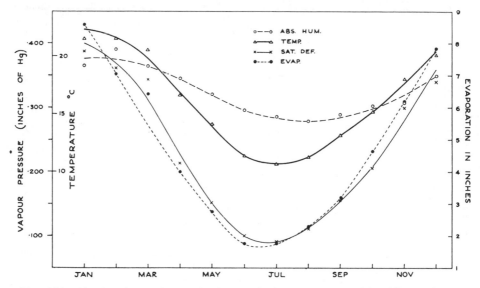

Fig. 6.04.—Showing the secular trends of atmospheric temperature and humidity at the Waite Institute, Adelaide. The curves are based on mean daily records for 23 years. Note that changes in absolute humidity are slight and gradual compared to the more abrupt changes in saturation deficiency and evaporation, which are closely related to temperature as well as to absolute humidity.

cated techniques are needed to evaluate this. The student who wishes to assimilate the "philosophy" behind this argument should read the section on *partial regression* in a good modern textbook on statistical methods. The immediate practical outcome is that causal relationships may be overlooked by concentrating on an obvious simple correlation; but the descriptive value of such calculations may remain unimpaired. Field observations are often related to temperature because it is more easily measured: Clark (1947a) got an adequate estimate of the temperature associated with the behavior of locusts by simply placing a thermometer with a blackened bulb among the insects in the situations that they occupied. And Wellington (1949a) painted the bimetallic strip of a thermograph black and exposed it in the foliage of spruce trees. Such calculations are valuable for descriptive purposes, provided that we remember that in this method temperature is being allowed to "speak for" humidity and any other variable in the environment which is correlated with temperature and that therefore the causal relationships may be quite other than what they seem to be at first sight.

The behavior of locust nymphs and adults in the field has been the subject of much study because it is generally accepted that high numbers in the outbreak area are not by themselves sufficient to produce the gregarious and migratory behavior which is such an essential feature of the destructiveness of these insects. Given high numbers, it is still necessary to have a stimulus which will bring about "aggregation" or crowding while the insects are in an active or excitable condition. Provided that the terrain is suitable, temperature may act in just that way, for the insects tend to crowd into local situations where the temperature approaches their "preferendum." With *Chortoicetes terminifera* this is about 42° C. (Clark, 1947*a*, 1949).

Clark studied the behavior of early-stage nymphs of this locust in an outbreak center in New South Wales. The first requirement of large numbers was satisfied in this case; for at the time when the eggs had just hatched, the first-instar nymphs were estimated to occur in the vicinity of the egg beds in numbers up to 2,000 per square yard in some situations. Notwithstanding the fact that their parents had been gregarious, the first-instar nymphs were neither markedly gregarious nor excitable. They soon distributed themselves over the area in relation to the occurrence of green plants suitable for food. Most of them spent the night sheltering in the vegetation or in other places. At daybreak the temperature in these situations was about 2° C., and the hoppers remained there motionless. They became more active as the temperature increased, until at 20° C. "disability" due to cold could no longer be detected and their movements were "normal." Between 20° and 45° C. the hoppers sought out and crowded into the warmest possible situations at ground level, taking up what the author calls the "basking formation." They crowded closely together, often with their bodies touching and broadside to the sun. On cool days basking groups persisted until late in the afternoon, usually breaking up as the shadows lengthened and the temperature fell. On hot days the basking groups would disperse as the temperature approached 45° C., and the hoppers would seek more shady places; ordinarily the groups would form again later in the afternoon. This behavior was governed by the existence of a "preferendum" at about 42° C. Outbursts of jumping would occur in the basking groups at irregular intervals, and as time went on the innate gregariousness and excitability of the hoppers increased. The first occurrence of gregarious mass migration was observed in the second instar in situations where the numbers were highest; elsewhere in situations where the numbers were lower, it was not observed until the fourth or fifth instar. The author summarizes his observations in the following words:

"The change from individualistic to gregarious behaviour is a result of increasing responsiveness of hoppers to the presence and movements of others.

It is effected by a period of crowding during which mutual contact, both visual and mechanical, becomes probably the most common experience of hoppers. External influences, especially temperature, play an important causal role in crowding. A period of crowding is apparently essential for the progeny of swarms of this species to *develop* gregarious behaviour, i.e., the capacity for mass migration. This is probably true for other locusts. The length of the necessary period of crowding varies in relation to population density and, probably, temperature conditions." Kennedy's (1939) account of the behavior of the desert locust *Schistocerca gregaria* in an outbreak center in the Sudan was in general agreement with the results described here for *Chortoicetes*. Kennedy concluded that "aggregation" of *Schistocerca* was largely the result of a "diurnal regime resembling in many ways that of phase *gregaria*, determined largely by temperature differences and changes, and involving prolonged basking on the small patches of bare sheltered and sunlit ground among the vegetation. As the patches are few in relation to number of locusts present, aggregation occurs."

Bogert (1952) observed similar behavior in certain lizards which live in Florida and Arizona. The lizards moved between sunny and shady places according to the temperature of their bodies. They were sensitive to small differences in temperature and usually were able, by this method, to keep the temperature of their bodies within a range of about $3°$ C. In certain sorts of weather the lizards would remain abroad all day, but on a clear summer's day, with the sun near its zenith, nearly all the lizards would be found sheltering in shady places. Bogert pointed out that lizards are able to carry on their "normal" activities only within a relatively narrow range of temperature. He suggested that this may explain why the distributions of large reptiles are restricted to the tropics. The large body, with its relatively small surface area, would take a long time to warm up. Bogert also pointed out that amphibians do not behave like this, nor do they seem to have "temperature preferenda."

We leave this subject now and pass on to another aspect of the influence of temperature on the activity of animals which, although it is of great importance in ecology, has been less studied than the matter of "preferenda."

6.22   *The Influence of Temperature on the Rate of Dispersal*

In chapter 5 we discussed the tendency that animals have to move about, leading to the dispersal of the population. In chapter 14 we give reasons for considering this sort of activity one of the most important characteristics of an animal, helping to determine the limits of its distribution and the level of its abundance, particularly when predators are important or food is sparsely distributed in the area where it lives. It is therefore important to understand how this activity may be influenced by the animal's environment. The several examples given in this section show how temperature may influence the dispersal-rate and indicate some of the ways in which this influence may be measured.

Perhaps the simplest approach to this problem is provided by the laboratory experiment in which temperature is carefully controlled. The speed of movement of the experimental animals is then observed directly, as the temperature in the place where they are living is varied at will. Such experiments do not, of course, supply any information about the behavior of animals living naturally, but they may provide valuable background for the interpretation of the more complex observations that may be made in nature.

The speed at which goldfish (*Carassius*) could swim at different temperatures was measured by Fry and Hart (1948). As was to be expected, the fish could swim more rapidly at high temperature than at low within the favorable range. But an interesting and important result was that the speed at which the fish could swim was determined not entirely by the present temperature, but it depended also on the temperature at which they had recently been living. In other words, the phenomenon of acclimatization, which is important in relation to the limits of the favorable zone (sec. 6.322), can also be observed in relation to activity within the favorable zone.

In these experiments a uniform group of young fish was divided into 7 lots and kept at 7 different temperatures between 5° and 35° C. until they had become fully acclimatized. They were then placed, one at a time, in a special rotating tank which stimulated them to keep swimming and in which it was possible to measure the speed at which they swam at any temperature. The maximal speed which the fish could maintain continuously for 2 minutes was measured and called the "cruising speed." When the cruising speed was measured at the same temperature as had been used for acclimatization, the maximal cruising speed was attained at about 28° C. But when the fish had been acclimatized at some temperature other than that at which the cruising speed was subsequently measured, it was found that the maximal cruising speed was always attained by fish which had been living near the temperature at which the test was made. The cruising speed was markedly less for fish that had been living at temperatures either much above or much below the temperature at which the test was made (Fig. 6.05).

It is well known that after the weevil *Calandra oryzae* has been breeding for a number of generations in a large bin of wheat, the insects will all be concentrated within a foot or two of the surface, irrespective of whether the colony started near the surface or in the depths of the bin. Birch (1946*b*) showed that this was due to the influence of temperature on the dispersal of the adult weevils. He placed small colonies of weevils on the surface and at various depths in several experimental bins of wheat. As the colonies developed in the depths, they caused the temperature of the wheat immediately around them to rise.

As the temperature reached and passed 32° C. (due to the heat of metabolism of the immobile young stages in the grain), the adults moved away to cooler places. Eventually, the layer of wheat near the surface was the only place which was cool enough to support the colony. These experiments showed that *C. oryzae* dispersed scarcely at all until the place where they were living became uncomfortably hot.

Another example of the direct approach to this problem is given by the observations of Gunn *et al.* (1948) on the behavior of migrating swarms of the

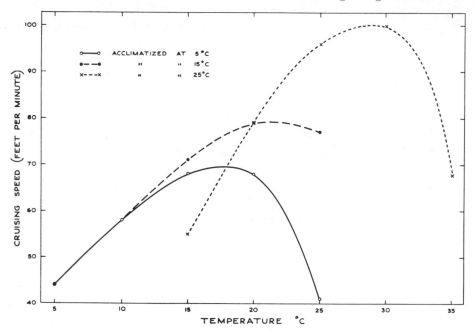

FIG. 6.05.—Showing the influence of acclimatization on the "cruising" speed of goldfish at various temperatures. Note how the temperature at which maximal "cruising" speed was attained was influenced by the temperature of acclimatization. (Data from Fry and Hart, 1948.)

locust *Schistocerca gregaria* in Kenya. Swarms of this locust often spend the night "roosting" in trees or shrubs. In the morning they descend to the ground. After spending about an hour on the ground basking in the sun, they usually take off on a gregarious migratory flight. A consistent routine is followed. At first, the insects merely bask in the sun; as the temperature rises, they begin to make short low flights, but at first there are very few in the air at any one time; later the numbers making flights increase, and each one stays aloft longer, so that, as time passes, more and more are seen in the air at once, flying over their comrades on the ground. Eventually enough are moving to stimulate the rest to take off more or less en masse. According to Gunn *et al.* (1948), the moment at which this mass flight begins is related to temperature but not to humidity or any other component of the environment which they measured. Swarms were occasionally observed to begin migrating when the temperature was as low as 14°–16° C.; once a swarm waited until the temperature had risen to 23°. But, as a rule, migration began when the temperature was between 17° and 22° C. Migration began at a lower temperature when the maximal temperature for the previous day had been low, and vice versa. In other words, the influence of temperature on the beginning of migration was modified by some stimulus which was correlated with the maximal temperature of the previous day. The explanation advanced by Rainey and Waloff (1951) is as follows. It is much easier for locusts to become "air-borne" and therefore to begin migrating when there are upward thermal currents near the surface of the ground. Therefore, migration usually does not begin until these currents develop or, in their absence, begins only at a relatively high temperature, when the locusts are more vigorous. Upward thermal movements usually develop with rising temperature in the morning, but the temperature at which they appear depends on the temperature of the air some distance above the ground. This is related to the maximal temperature of the previous day. In this way the correlation between the temperature at which migration begins and the maximal temperature of the previous day is explained. Provided that the temperature of the air exceeds a certain threshold (about 16° C.), the swarm will begin migrating just as soon as the thermal movements of the air are sufficient to enable them to become "air-borne." In the absence of favorable air currents, the locusts may still become "air-borne," but this requires much more energy on their part and does not happen until the temperature approaches 22°–23° C. Confirmatory evidence for this explanation of the data is given by the complementary observation that in the afternoon migratory swarms usually cease flying about the time that the temperature falls to 19°–23° C. It is of interest to recall that Kennedy (1939) reported that the solitary phase of this locust was stimulated by light to take off on its dispersal flight, provided that the temperature exceeded a certain threshold (sec. 5.2).

With other species it may be impracticable to make direct observations of this sort. But with suitable experiments, using the appropriate statistical methods, it is still possible to make precise inferences about the influence of temperature on the dispersive behavior of these species. A method which has the merit of being direct and simple is exemplified by the experiments of Dobzhansky and Wright (1943). Between 3,000 and 5,000 marked flies (*Drosophila pseudoobscura*) were liberated in an area where a natural population of the same species was already living at an average density of from 4 to 8 flies to each 100 square meters of territory. In each of four experiments the marked flies were liberated all together at the same place. Traps were then set out, spaced 20 meters apart and arranged in the form of a square cross with the center of the cross at the place where the flies had been liberated. During the next 5–10 days the flies in each trap were recorded daily and then set free again at the same place where they had been caught. The results for any one day gave the distribution of the flies on that day, and a comparison of the distributions on successive days gave information about the behavior of the flies—their rate of dispersal and so on. Figure 6.06 illustrates the distribution of the marked flies on two successive days in one experiment. One characteristic of the distribution is its variance, $\Sigma fr^2/\Sigma f$, where Σ has its usual meaning as the symbol of

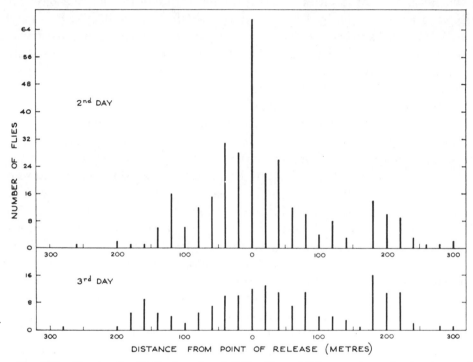

FIG. 6.06.—Showing the distribution of marked flies (*Drosophila*) on two successive days in one of Dobzhansky's experiments. See text for explanation. The unit for the abscissae in this diagram is 20 meters—the distance between traps; the ordinates show the number of flies in each trap. (After Dobzhansky and Wright, 1943.)

summation, f is the frequency (i.e., the number of flies in the trap), and r is the distance of the trap from the origin of the distribution.

If the variance at the end of the first day was s^2 and then, if the flies continue to disperse at the same rate, the variance at the end of the second day will be $2s^2$ and at the end of the tth day ts^2. On the other hand, if the rate of dispersal falls off with time (as the flies become less crowded, for example), then the daily increment in variance will be less than s^2, and the variance of the distribution on the tth day will be less than ts^2. In these experiments (as can be seen from Table 6.01) the rate of dispersal did not fall off with time; on the last (tth) day of each experiment the variance was about t times greater than it had been on the first day. This means that the variance increased, on the average, by about s^2 each day; but Table 6.01 also shows that the increase was far from uniform from day to day. A large part of this variability could be attributed to the influence of temperature on the activity of the flies.

When the daily increments in variance were correlated with daily temperature, the coefficient for linear regression of variance on temperature was found to be 2.31 ± 0.50. This means that, on the average, the daily increase in variance increased by 2.31 units (400 square meters) for each increase of 1° F. The regression was not truly linear, for the rate of dispersal was relatively slow below 20° C. and much faster above this temperature. The general conclusions to be drawn from these experiments, which are relevant to this section, are:

TABLE 6.01*
VARIANCES OF DISTRIBUTIONS OF MARKED FLIES ON SUCCESSIVE DAYS AFTER LIBERATION

Day	I	II	III	II¹	III¹	IV¹
1 (s^2_1)	3.8 (279)	8.7 (584)	8.4 (674)	7.6 (337)	9.0 (295)	11.3 (635)
2	8.1 (609)	21.0 (354)	22.2 (532)	22.5 (220)	30.5 (236)	22.8 (306)
3	13.6 (369)	25.8 (238)	42.6 (228)	22.7 (128)	52.2 (133)	46.5 (102)
4	13.0 (171)	34.5 (276)	41.6 (166)	35.0 (168)	40.3 (69)	78.4 (73)
5	10.4 (94)	37.5 (145)	......	35.7 (90)	44.3 (123)	73.1 (51)
6	17.8 (215)	......	......	63.2 (78)	73.0 (58)	......
7	21.4 (81)	......	......	64.5 (79)	90.4 (39)	......
8	32.4 (69)	......	......	......	......	......
ts^2_1	30.4	43.5	33.6	53.2	63.0	56.5

* Experiments I, II, and III based on traps arranged along 4 radii at 90°. Experiments II¹, III¹, and IV¹ based on traps along 2 radii at 180°. This arm of the original cross was extended by adding traps to each end after the flies had reached the perimeter of the original area. The figures in parenthesis give the number of flies caught that day (Σf). After Dobzhansky and Wright (1943).

(a) apart from the first day, when some 3,000–5,000 flies were liberated at one place, the density of the population had no appreciable influence on the rate of dispersal; (b) the temperature exerted a significant influence on the rate of dispersal, and, in particular, this increased with temperature when the temperature exceeded 20° C. Certain other inferences about the behavior of *Drosophila* in a natural population may be drawn from the results of these most interesting experiments, and these have been discussed in the appropriate places elsewhere (sec. 5.11).

When the raw data to be analyzed are derived, not from an experiment of the sort that was done with *Drosophila*, but rather from observations made on an undisturbed natural population, then other methods of analysis may be needed. The statistical device of partial regression is often useful. As an example we may describe the quantitative study of the movements of the flower-inhabiting thrips (*Thrips imaginis*) made by Davidson and Andrewartha (1948a, b). This species is strongly attracted to roses, and these flowers were used as traps. The insects were breeding in a variety of other flowers in and around the garden, and these were the reservoir from which the thrips came to the freshly opened roses. Daily records of the numbers of thrips in roses were kept for the months August–December (southern spring) for 14 years. A characteristic series of records is shown in Figure 6.07, which represents the daily numbers of thrips in roses during 1932.

The seasonal trend and the enormous daily fluctuations about the trend shown for this year are quite typical for all the other years for which records were kept. The first step in analyzing the data is to transform the numbers to logarithms: this is done chiefly because proportional changes in numbers are more instructive than additive changes. It is then possible to account for the

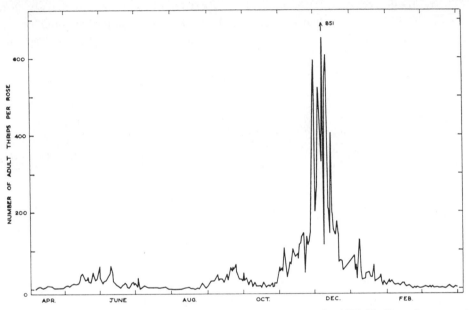

FIG. 6.07.—Showing the daily mean numbers of thrips per rose for 1932–33. Note the seasonal trend and the marked daily fluctuations about the trend. (After Davidson and Andrewartha, 1948a.)

systematic trend of numbers with time by fitting empirically a curvilinear regression of the form

$$Y = a + bx + cx^2 + dx^3 ,$$

where Y represents the logarithm of the number of thrips observed and x represents days. A typical curve is illustrated in Figure 6.08. The systematic trend which this curve describes is due chiefly to the natural increase of the population with time, but it may include other components associated with systematic trends in the weather. The important point to note is that the curve has accounted for the trend with time, whatever may be its cause, so that the residual variance (represented by the departures of the observed points from the trend line in Fig. 6.08) is independent of time. These departures are partly (probably largely) due to daily variations in the degree of activity displayed by the thrips in moving into and out of the "trap" flowers. On a warm day more thrips will move out of the flowers in the surrounding fields, where they have been breeding, and find their way into the newly opened roses, which are sampled daily. It is possible to evaluate precisely the degree of association between the "activity" of the thrips (as represented by the residual departures from the trend line) and the various components of environment which, experience suggests,

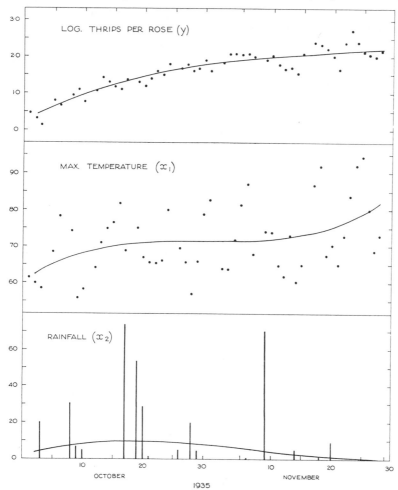

Fig. 6.08.—Showing (*a*) daily records of the number of thrips per rose during the spring of 1935 (the numbers have been expressed as logarithms); (*b*) the daily maximal temperature for the same period in degrees Fahrenheit; and (*c*) the daily rainfall. The unit is one-hundredth of an inch. The trend was in each case obtained by calculating the appropriate polynomial of the form $Y = a + bx + cx^2 + dx^3$. (After Davidson and Andrewartha, 1948*b*.)

may be important. The final step in the analysis is to solve the equation for partial regression, which may be written,

$$Y = b_1 x_1 + b_2 x_2 + b_3 x_3 + \ldots ,$$

where Y represents the departure of the observed data from the trend line in Figure 6.08; $x_1, x_2, x_3, \ldots$, represent temperature, humidity, and other components of environment, also expressed as departures from the trend line with time; and b_1, b_2, b_3, . . . , are the partial regression coefficients such that b_i measures the association of y with x_i, independent of any relationships which might exist between these and any other variable in the equation.

In the present example it was found that the independent influence of temperature was highly significant and that, on the average, the numbers of thrips in the flowers on any one day increased by 25 per cent for each increase of 5° C. in the daily maximal atmospheric temperature. The only other component of environment which was found to be significantly related to the activity of the thrips was daily rainfall. But this was less important than temperature. In the particular circumstances of this investigation the measurable activity of the thrips which could be attributed to temperature was about three times that associated with rainfall.

A striking and most important feature of the data illustrated in Figure 6.08 is that, for any given fluctuation in temperature, the relative increase or decrease in the numbers of thrips (i.e., the magnitude of the departures from the trend line) is about the same, whether the absolute numbers of thrips are low or high. This indicates that the movements of the thrips to and from the flowers is not dependent on density; it is not the outcome of jostling or any other manifestation of numbers. But it could be the outcome of an innate "wanderlust," possessed to an equal degree by thrips living in solitude and those living in a crowd. This tendency may be modified by several components of the environment, of which temperature has been shown to be an important one.

Williams (1939, 1940) was the first to apply the method of partial regression to get a precise measure of the independent influence of different components of the environment on the dispersal of insects in nature. He counted the numbers of insects of all kinds caught each night in a light trap that ran continuously for 4 years at Rothamsted. He related these numbers to the daily records of temperature, humidity, moonlight, and so on. The results indicate that relatively more insects came into the trap in warm weather than in cold and that none of the other components of the environment that were measured influenced the counts as much as temperature. These two papers are especially valuable for their full discussion of method. We have not used them for our main example in this section because, with so many species being considered together, the interpretation is not so clear as in the simpler case with *Thrips imaginis.*

6.23   *The Influence of Temperature on the Speed of Development*

It has long been known that poikilothermic animals complete their development more rapidly in warm weather than in cool. The speed of development at different constant temperatures has been measured for many species, and several mathematical expressions have been proposed which purport to describe the relationship of speed of development to temperature. For technical reasons it is inevitable that constant temperatures are used in such experiments.

6.231 CONSTANT TEMPERATURE

Figure 6.09 illustrates two alternative ways that are often used to represent the speed of development at different temperatures. The temperature is plotted as the abscissae, and the ordinates may be either the duration of the particular

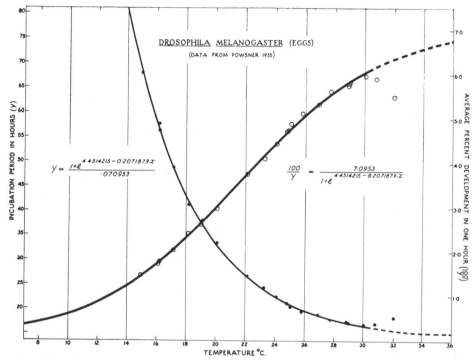

FIG. 6.09.—Showing the speed of development of eggs of *Drosophila melanogaster*. The hours required to complete the egg stage are plotted against temperature (*descending curve*), and the reciprocal of this time is also plotted against temperature (*ascending curve*). The points plotted on the diagram are observed values in each case. The ascending logistic curve was fitted to the reciprocals of the observed times. The calculated values for the descending curve were the reciprocals of the calculated values for the logistic curve. (After Davidson, 1944.)

stage that is being studied or, more often, the reciprocal of this measurement. Commonly, the latter is multiplied by 100, so that it becomes per cent development per unit time (day or hour). The transformed data are the usual starting point for generalized mathematical expressions or curves. But note that the reciprocal is a highly specialized conception of speed of development and has a number of peculiar properties. For example, the slope of the curve is a function of the duration of the particular stage under consideration. This becomes immediately apparent by a simple analogy. Two vehicles travel at identical constant speed along a straight road. One has twice as far to go as the other. If the "speed" of the cars be estimated as a percentage of the journey completed in an hour, one will be judged to have twice the "speed" of the other. This peculiar property of the reciprocal scale makes it valueless for comparisons either between the same stage in different animals or between different stages in the same animal.

Fry (1947) examined a number of the mathematical expressions which have been proposed to describe the relationship of temperature to speed of development. He classified these into "physiological" or "biological," according to the purpose for which they were intended. He designated the expressions of vant'Hoff, Arrhenius, and Bělehrádek as "physiological" and that of Janisch and the well-known method of "thermal sums" as "biological." Davidson (1944) called the equations of vant'Hoff and Arrhenius "theoretical" expressions, and the others, including the logistic curve which he was himself proposing, "empirical." We think this is the better classification.

The Arrhenius equation is usually written as follows:

$$\frac{Y_2}{Y_1} = e^{\frac{1}{2}\mu\left(\frac{1}{x_1} - \frac{1}{x_2}\right)},$$

where Y represents speed of development at temperature x measured in degrees absolute, and μ is a constant. The vant'Hoff equation is usually written:

$$Q_{10} = \left(\frac{Y_1}{Y_2}\right)^{\left(\frac{10}{x_1 - x_2}\right)},$$

where Y represents the speed of development and x represents temperature on the centigrade scale, and Q_{10} is a coefficient. These expressions may be written:

$$\mu = 4.6(\log Y_2 - \log Y_1) \times \frac{1}{(1/x_1) - (1/x_2)}$$

and

$$\log Q_{10} = 10(\log Y_1 - \log Y_2) \times \frac{1}{x_1 - x_2}.$$

By hypothesis, μ and Q_{10} are constant. When the equations are written in this form, it becomes clear that a constant value for μ implies a linear relationship between the logarithm of the speed of development and the reciprocal of temperature on the absolute scale; and a constant value for Q_{10} implies a linear relationship between the logarithm of the speed of development and temperature on the centigrade scale. But Bělehrádek (1935) pointed out that the reciprocal of the absolute temperature is practically a linear function of temperature on the centigrade scale between the limits 0° and 40° C. So that the expressions of Arrhenius and vant'Hoff are virtually equivalent; both imply that the proportional increase in speed of development produced by a given difference in temperature is constant throughout the temperature range at which an animal may develop. If μ fits any particular set of empirical data, then Q_{10} should fit equally well, and vice versa. In practice, it has been shown repeatedly that Q_{10} and μ are far from constant and that they vary in a systematic way with temperature. Bělehrádek (1935) quoted numerous authors who had found either μ or Q_{10} to be inconstant. We shall illustrate the inconstancy of μ with just one example taken from the work of Bliss (1926). Working with the prepupal stage of *Drosophila*, he found the following values for μ:

Between 12° and 16° C. $\mu = 33,210$
Between 16° and 25° C. $\mu = 16,800$
Between 25° and 30° C. $\mu = 7,100$

This is but a quantitative example of the principle which Barcroft (in Fry, 1947) recognized when he pointed out that over a range of temperature for which the animal is best suited the temperature-metabolism curve is relatively flat. For example, between 6° and 20° C., the standard metabolism of the trout is doubled but that of the goldfish is increased fivefold. This is about the temperature range that best suits the trout, but the goldfish is adapted to much higher temperatures (Fry, 1947). When we consider that Q_{10} and μ were originally designed to describe the relationship of temperature to a single chemical reaction, it is hardly surprising that they should be inadequate when they are related to the immensely complex chain of reactions which proceed during the morphogenesis of an animal. These coefficients have also been applied, seemingly indiscriminately, to measurements of rate of activity, such as heartbeat, rate of nervous conduction, velocity of locomotion (*Paramecium*), and so on. These are much simpler processes than development, and it may be that a simpler expression will describe them. But we are not immediately concerned with this usage, except to point out that there is no good reason for expecting measurements of activity to be equivalent to measurements of metabolism. In view of the chemical complexities of even the simplest morphogenesis, it seems unlikely that a simple (or, indeed, any) theoretical expression will be found that adequately relates temperature to the speed of development. In the meantime, the purposes of ecology are best served by seeking a satisfactory empirical equation to describe this relationship.

The best known of the empirical equations is the hyperbola:

$$y(x - a) = K.$$

In this equation y is the duration of development, usually expressed in days or hours, at temperature x, which is usually expressed in degrees centigrade.

When y is converted to its reciprocal, which is the conventional measure for speed of development, this equation becomes:

$$\frac{1}{y} = k + bx,$$

where y and x have the same values as before and k is a new constant equal to $-ab$. This is the equation of a straight line, and a can be shown to be the value of x where the value of y becomes zero, i.e., the temperature at which the "speed of development" becomes zero. It has been called the "threshold of development." The supposed linear relationship between temperature and the speed of development has been widely accepted in the past (sec. 6.234). However, it is now well known, and any well-conducted experiment will demonstrate, that the temperature-time curve approximates to the hyperbola for only a very short range of temperature, departing widely from it at either end of the favorable range. Moreover, it is never possible to get empirical points near the theoretical "threshold," so that the value of a is determined by extrapolation—a statistical procedure which in this instance has virtually no biological meaning.

Realizing that the hyperbola is inadequate, Bělehrádek (1935 and earlier papers) proposed the equation

$$Y = \frac{a}{x^b},$$

where Y is time required to complete development at temperature x, which is usually expressed in degrees centigrade, and a and b are constants. This is the equation of an exponential expression which becomes linear on the logarithmic scale, taking the form

$$\log Y = \log a - b \log x,$$

and b is seen to be the coefficient of linear regression of $\log y$ on $\log x$. In this abstract sense b has the meaning of a temperature coefficient; but with both axes of the graph on the logarithmic scale, it is difficult to attribute any real meaning to b. However, since this is an empirical equation, the proper criterion for its usefulness is how well it follows the empirical data. It will not be sufficient just to rely on visual inspection, for it is well known that the double-log scale greatly minimizes the apparent discrepancies from the curve, though these remain to be demonstrated by the appropriate statistical methods. But even by visual inspection the empirical equation of Bělehrádek can be shown to be inferior to the logistic expression which we describe below (Fig. 6.10).

The logistic equation was derived by Verhulst, and rediscovered by Pearl (Pearl and Reed, 1920), to describe the trend of increase in a population growing where the space was limited (sec. 9.21). In this context the equation has a theoretical basis; but Davidson, having observed the consistent recurrence of a sigmoid trend in temperature-development data, fitted the Pearl-Verhulst logistic curve as a purely empirical description of this trend. He searched the literature for experiments that appeared to have been carried out with the necessary precision. He found that in every case the logistic curve followed the observed data more closely than did any of the older expressions. He published

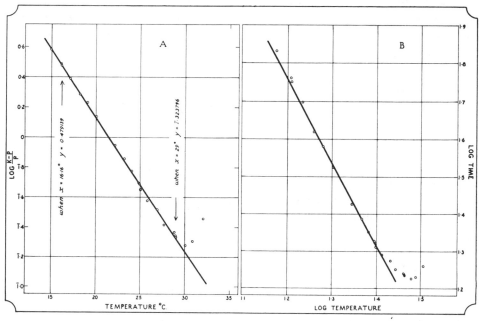

FIG. 6.10.—Showing *A*, the linear transformation of the logistic curve; *B*, the straight line derived from the Bělehrádek equation for the same data for *Drosophila*. Note that the former gives a closer fit over a greater proportion of the temperature range at which development may proceed. (After Davidson, 1944.)

his findings in two brief papers (Davidson, 1942, 1944), which should be consulted in the original, for they constitute a completely adequate exposition of the use of the logistic equation to relate temperature to speed of development of insects. The logistic curve is realistic, giving an easily comprehended picture of the trend of speed of development at different constant temperatures. It is easily calculated directly from the empirical data, and, in addition, it has the merit of being the most adequate empirical description of this relationship that has so far been suggested. From now on, it should be used in preference to all the older expressions.

The logistic equation may be written

$$\frac{1}{Y} = \frac{K}{1 + e^{a-bx}},$$

where Y represents the time required to complete development at temperature x, which is usually measured in degrees centigrade, and a, b, and K are constants. K defines the upper asymptote toward which the curve is trending; b defines the slope of the curve; and a relative to b fixes its position along the x axis. This may be the more easily comprehended by considering that the logistic curve is a bi-symmetrical sigmoid curve having a point of inflection (the steepest point on the curve) whose co-ordinates are $1/Y = K/2$ and $x = a/b$. The ordinate $1/Y$, being the reciprocal of the time required to complete a particular stage, measures the proportion of the total development completed in unit time; it thus serves as a conventional measure of speed of development. In practice, it is convenient to multiply this by 100 and express time in days (or hours), so that the right-hand side of the equation expresses per cent development per day (or per hour).

The constants K, a, and b are readily calculated from the data, making use of the transformation

$$\log_e \frac{K - y}{y} = a - bx.$$

This is the equation of a straight line. Given the value of K and any two values of y corresponding to particular values of x, the constants a and b are readily computed. The usual method is to make a first approximation for K and two values of y from a freehand graph, then, using the linear transformation given above, to improve these values by several successive approximations. The whole process of fitting may be carried out speedily and with quite adequate accuracy by using graphical methods throughout (Pearl, 1930, p. 420; Davidson, 1944). Alternatively, once K has been fixed by interpolation, the constants a and b may be calculated by the usual method of linear regression (Davidson, 1944).

Applying this method to Powsner's (1935) data for the eggs of *Drosophila*, Davidson (1944) obtained the equation

$$\frac{100}{Y} = \frac{7.0953}{1 + e^{4.451422 - 0.207188x}}.$$

The resulting curve is shown in Figure 6.09. It will be seen that the observed points cluster quite closely about the calculated curve between 15° and 30° C. Above 30° the harmful influence of high temperature is manifest from the pronounced decrease in the speed of development above that temperature. The reciprocal of the logistic is also shown in Figure 6.09. This is calculated directly from the logistic by taking the reciprocals of the calculated values. It transforms the data back into their original form, namely, time required to complete development. It seems less instructive than the simple logistic curve, but it may be useful in certain *ad hoc* calculations. Figure 6.10, *A*, shows the straight line obtained by plotting log $(K - y)/y$ against temperature; and Figure 6.10, *B*, is included so that the logistic may be compared with Bĕlehrádek's equation. The superiority of the former is evident.

So far, the logistic equation has been applied to data for the speed of development of insects (Davidson, 1942, 1943*a*, *b*, 1944; Birch, 1944*a*, 1945*c*;

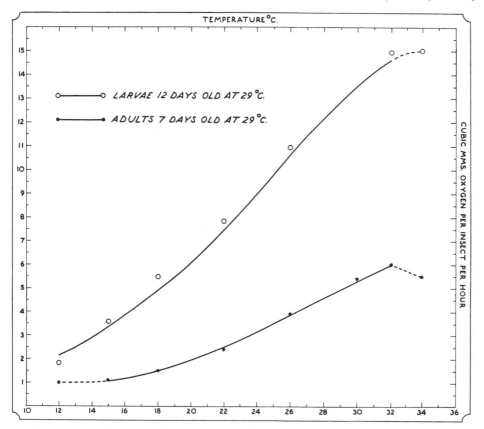

FIG. 6.11.—Showing the relationship between temperature and the rate of oxygen consumption by larval and adult stages of *Calandra oryzae*. (After Birch, 1947.)

Birch and Snowball, 1945; Browning, 1952*b*). But Birch (1947) found that it also was appropriate for data relating the rate of oxygen uptake to temperature in developing larvae of *Rhizopertha dominica* and *Calandra oryzae* (Fig. 6.11).

It was pointed out above that the logistic equation gives a curve which has a point of inflection at the co-ordinates $1/Y = K/2$, $x = a/b$; that is, the steepness of the curve increases up to this point and then gradually decreases. This means that for any given increment in temperature the increase in the speed of development is greatest for median temperatures and gradually falls off at higher temperatures. The fact that the acceleration fell off after the point of inflection led Davidson (1944) to suggest that the unfavorable influence of high temperature may be beginning to make itself increasingly evident from this point onward. But this is a theoretical consideration which has little bearing on the usefulness of the logistic equation as an empirical description of the relationship between temperature and the speed of development of insects. There are a number of considerations which make it desirable to refrain from imputing a theoretical foundation for the logistic equation.

It has been shown that the χ^2 test may be used to test the goodness of fit of the logistic curve (Birch, 1944*a;* Browning, 1952*b*), provided that the final approximations of the constants K, a, and b are found by solving the appropriate equations for maximal likelihood. The χ^2 test essentially measures the magnitude of the departures of the observed values from the hypothetical ones relative to the intrinsic variability of the material being observed. The precision of the test is enhanced by increasing the number of replicates at each temperature.

Browning (1952*b*) applied the χ^2 test to his own data for eggs of *Gryllulus*, to those of Birch (1944) for eggs of *Calandra*, and to those of Powsner (1935) for eggs of *Drosophila*. The last are those used by Davidson (1944) and reproduced in Figure 6.09. These three sets of data were chosen because in each case they had been derived from adequately large samples (total exceeded 500 in each case) and the experimental work had been done precisely. The values for χ^2 for these three sets of data are set out in Table 6.02. The relevant curves are

TABLE 6.02*
GOODNESS OF FIT OF CALCULATED CURVES SHOWN IN FIGURES 6.09, 6.12, AND 6.13

Source of Data	χ^2	Degrees of Freedom	P
Eggs of *Drosophila* (Powsner).........	1,053.6	17	<0.0000
Eggs of *Calandra* (Birch)............	29.6	2	<0.0000
Eggs of *Gryllulus* (Browning).........	82.6	3	<0.0000

* After Browning (1952*b*).

shown in Figures 6.09, 6.12, and 6.13; from these figures it may be seen that in each case the departures of the observed points from the hypothetical curve are small indeed. Yet the final column in Table 6.02 indicates that in each case discrepancies as large as these would not have occurred by chance once in many thousands of trials. This means that there is a significant proportion of the variance due to differences in temperature left over after the logistic curve has accounted for its portion. Or, in more colloquial terms, we might say that in each of the three cases we have been able to show that there are odds of very much better than 10,000 to 1 against the logistic curve's being the "true" representation of the relationship between temperature and the speed of development.

There are sound theoretical reasons for expecting just this result. Fortunately, Powsner (1935) published not only the data for the egg stage of *Drosophila* which we have been discussing but also equally precise data for the larval and pupal stages. Browning (1952*b*) re analyzed these data first for each stage separately and then by adding together the time required for egg and

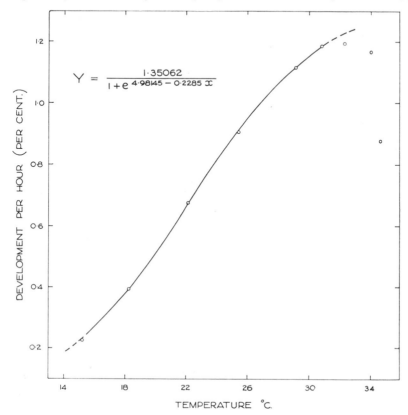

$$Y = \frac{1.35062}{1 + e^{\,4.98145 - 0.2285\,x}}$$

FIG. 6.12.—Showing the relationship between temperature and the speed of development of the egg stage of *Calandra oryzae*. (After Birch, 1944*a*.)

larval stages, and, finally, for egg, larval, and pupal stages taken together (Fig. 6.14; Table 6.03). Two striking results stand out. First, the curves

TABLE 6.03*
GOODNESS OF FIT OF CALCULATED CURVES SHOWN IN
FIGURE 6.14

Stage in Life-Cycle of *Drosophila*	x^2	Degrees of Freedom
Egg......................	1,053	17
Egg + larva...............	4,952	18
Egg + larva + pupa........	66,581	10

* After Browning (1952*b*).

for the different individual stages differ not only in the magnitude of the con-stants *a* and *b* but also in the range of temperature they occupy. Second, the best fit was for the egg stage, the poorest for the combined egg, larval, and pupal stages. The second of these results could have been predicted from the first. For even if the relationship between temperature and speed of development for the egg stage were truly expressed by the logistic equation and simi-

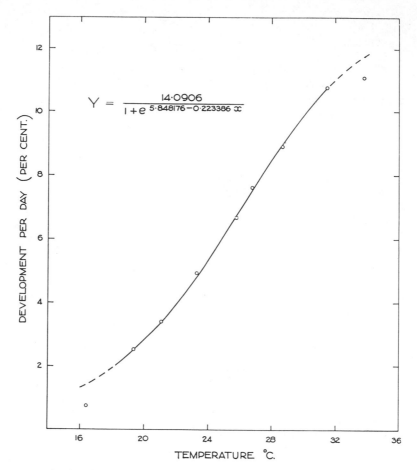

$$Y = \frac{14 \cdot 0906}{1 + e^{\,5 \cdot 848176 - 0 \cdot 223386\, x}}$$

FIG. 6.13.—Showing the relationship of temperature to the speed of development of the egg stage of *Gryllulus commodus*. (After Browning, 1952b.)

larly for the larval stage, still the two stages added together could conform to the logistic equation only if the constants a and b were the same in each of the two primary curves. In this case a and b are demonstrably different for the egg and larval stages (Fig. 6.14). A moment's reflection will indicate that even a seemingly uniform stage like embryogenesis is hardly likely to consist either of one uniform process or even of a series of successive processes so similar that the curves describing them will have the same constants a and b. It is therefore unlikely that any but the shortest and simplest stage of morphogenesis will be truly represented by the logistic equation; and even this still remains to be demonstrated.

These considerations, important as they may be to the theory of the subject, need not detract from the usefulness of the logistic curve as an empirical description of the relationship between temperature and the rate of development of poikilothermic animals. Figures 6.09, 6.12, and 6.13 show that the

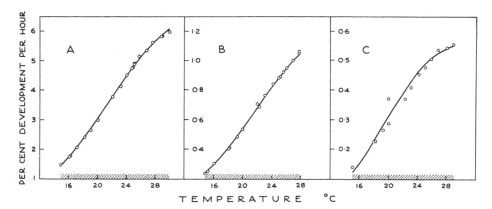

FIG. 6.14.—Showing the relationship between temperature and the speed of development of *A*, the egg stage; *B*, the egg and larval stages taken together; and *C*, the egg, larval, and pupal stages taken together, of *Drosophila melanogaster*. Note that the deviations from the trend line become progressively greater from *A* to *C*. (After Browning, 1952*b*.)

departures of the observed points from the trend lines are small in each case; and the calculated trend can be taken as adequate for most ecological purposes.

6.232 FLUCTUATING TEMPERATURES

In nature, animals live in places where the temperature fluctuates; so it is necessary to see how far the conclusions reached in the preceding section, which were based on experiments done at constant temperatures, may be applied to animals living in nature. There is no simple answer to this question. The matter is complicated by the following considerations: (*a*) The life-cycle may include a diapause-stage which needs to be completed at a lower temperature than that which favors the more obvious processes of morphogenesis (chap. 4). (*b*) The fluctuations may include extreme temperatures which are harmful, and even short exposures to these may impair the animal's competence to develop healthily at a favorable temperature. (*c*) Different stages in the life-cycle may have different limits to the favorable range and may respond differently to temperature within the favorable range. (*d*) Healthy development may proceed during short or intermittent exposures to extreme temperatures which would be harmful or even lethal if experienced continuously. (*e*) The relationship between temperature and the speed of development is not linear. We shall examine each of these five phenomena in the order in which they have been listed, beginning with diapause.

a) One of the most characteristic features of diapause is that it disappears during adequate exposure to "low" temperature (chap. 4). Diapause-development (that is, the physiological processes which culminate in the disappearance of diapause) proceeds most rapidly, as a rule, over a range of temperature that is lower than the favorable range for morphogenesis, although the two ranges

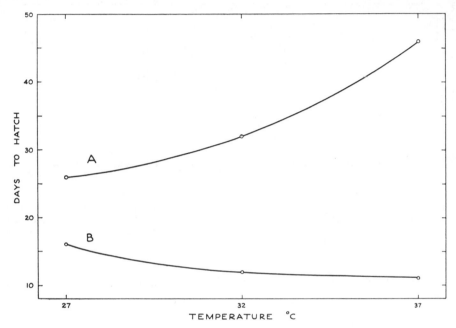

Fig. 6.15.—Showing the duration of development of the eggs of *Melanoplus mexicanus* at constant temperature. *A*, without any exposure to cold; and *B*, after 240 days at 0° C. Note that the slope of curve *B* is characteristic of "normal" nondiapause insects; *A* slopes in the opposite direction, because the weak diapause is somewhat "firmer" at 37° than at 27° C. (Data from Parker, 1930.)

usually overlap (Fig. 4.03). In cases where a clear-cut and firm diapause is present, no difficulties of interpretation need arise, and investigation will usually reveal that a specific, and often prolonged, exposure to low temperature is required before morphogenesis may proceed healthily or at all. But it was made clear in section 4.3 that diapause is not a clear-cut, all-or-none type of phenomenon. There are innumerable instances in nature of weak or incipient diapause and still others, like *Austroicetes*, which may be found at certain times of the year with the remnants of what was originally a firm diapause (sec. 4.1). In these cases the animals (or in marginal cases, where the diapause is more pronounced, a proportion of the animals) are competent to complete development at constant temperatures within the favorable range, without first experiencing cold, and the presence of diapause is thus not completely self-evident. But development is markedly more healthy in the lower part of the temperature range, and characteristically the increase in speed of development with increasing temperature is relatively small or may even be negative (Figs. 6.15, 6.16). Exposure to fluctuating temperature with the lower temperature in the range favorable for diapause-development or, alternatively, a single prolonged exposure to such a temperature usually results in a pronounced acceleration in the speed of development at higher temperatures (Fig. 6.16, *B*).

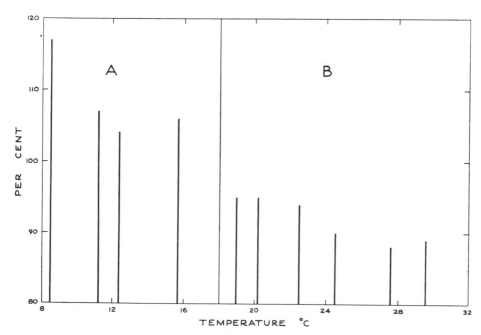

FIG. 6.16.—Showing the relative time required to complete development by "control" and "experimental" eggs of *Rana pipiens*. The ordinates show the time required for the "experimental" eggs to complete development, expressed as a percentage of the time required by the "control" eggs at the same temperature. *A:* All eggs completed the first part of their development at some temperature above 18° C. *B:* All eggs completed the early part of their development at some temperature below 18° C. Note that a weak or incipient diapause retarded the development of eggs represented in *A*, whereas the disappearance of this diapause from eggs represented in *B* caused the "experimental" eggs to develop more rapidly than the "controls." (Data from Ryan, 1941.)

But it is not correct to attribute this acceleration either to the "stimulating" influence of fluctuating temperatures or to acclimatization. It is more instructive to recognize it for what it is, namely, a manifestation of diapause. The eggs of *Melanoplus mexicanus* will serve as an example in which diapause is relatively well developed (Parker, 1930). The pupa of *Calliphora erythrocephala*, the eggs of *Locusta migratoria*, and the eggs of *Rana pipiens* are species in which diapause, as it is usually understood, does not occur. But with carefully planned experiments, making proper use of the appropriate statistical techniques, it is possible to demonstrate a small response to low temperature which is best interpreted as a "diapause-effect," i.e., it is a manifestation of a very weak or incipient diapause (Ahmad, 1936; Ryan, 1941). These four examples will now be discussed briefly.

The well-known work of Parker (1930) with eggs of *Melanoplus mexicanus* has often been quoted in textbooks and elsewhere as an example of the "stimulating" influence of low temperature. This is unfortunate, for it is indubitably an example of a weak diapause. A proportion of the eggs hatched without any exposure to low temperature; but the response to increasing temperature was quite the reverse of that usually found, the incubation period being longer, the higher the temperature. This response was reversed after a prolonged exposure to low temperature; not only was the usual increase in speed of development with increasing temperature observed, but the actual time at each temperature was also shortened (Fig. 6.15).

The pupae of *Calliphora erythrocephala* are apparently quite normal in their response to temperature in the favorable range: the data in Table 6.04, second column, would give no reason to suspect the presence of any sort of diapause. Yet when pupae were exposed to 5° C. for 8 days, the time required to complete development was reduced by about 10 per cent at each temperature except 30° (Table 6.04, third col.). The duration of the pupal period at 18° and 23° was also reduced by about the equivalent amount when pupae were exposed on alternate days to 5° and 18° or 23° (Ahmad, 1936). Ahmad considered

TABLE 6.04*
DURATION OF PUPAL PERIOD OF *Calliphora* IN DAYS AT DIFFERENT CONSTANT TEMPERATURES AND REDUCTION IN DURATION OF PUPAL PERIOD DUE TO PRELIMINARY EXPOSURE OF 8 DAYS AT 5° C.

Temperature (° C.)	Duration of Pupal Period (No Exposure to Cold) (Days)	Reduction in Pupal Period Due to Exposure of 8 Days at 5° C. (Days)
14.8	22.1	2.1
18.4	14.6	1.5
23.0	10.9	0.9
27.0	8.8	0.5
30.0	7.8	0.1

* After Ahmad (1936).

whether these results might be due to morphogenesis going on at 5° C., but rejected this possibility. There seems little doubt that here we are concerned with the same phenomenon as in the egg of *Melanoplus*, the difference being chiefly one of degree. In *Melanoplus* the diapause, though weak, is prominent enough to be recognized as a true diapause; in *Calliphora* a true diapause as such does not occur, yet the beginnings of an incipient diapause must be there, for the responses to constant temperature differ significantly from those to fluctuating temperature and in a way that is different from *Melanoplus* only in degree. In the eggs of *Locusta migratoria* the diapause-effect is even weaker than in pupae of *C. erythrocephala*, for maximal response was obtained after 1 day's exposure to 5° C.; in some circumstances, for example, with eggs that had completed 25 per cent of their development, an exposure of 4 days at 5° C. also resulted in a small but significant decrease in the time required to complete development. But continuous exposure for longer periods to 5° C. was harmful and apparently impaired the eggs' competence to develop healthily when they were returned to a favorable temperature (Ahmad, 1936).

Ryan (1941) incubated eggs of the frog *Rana pipiens* at a series of constant temperatures. "Control" batches remained at the same temperature through-out the experiment, but "experimental" batches came to these same tempera-tures for the latter part of their development after having completed the earlier part at either a higher or a lower temperature. The period required by the "experimental" eggs to complete a particular stage in their embryogenesis was then compared with the period required by the "control" eggs to complete the same stage. In Figure 6.16 the period required by the "experimental" eggs is expressed as a percentage of the period taken by the "control" eggs for the same stage at the same temperature (indicated on the abscissae). The results were typical of a "weak" or incipient diapause. Eggs which had completed their early development at a high temperature and were then placed at a "low" temperature developed more slowly than did the controls, which had been at the low temperature all the time. This could be interpreted as a disappearance of "diapause" from the "controls" at low temperature. When the transfer was made in the opposite direction, Ryan found that "experimental" eggs which had completed the first part of their development at "low" temperature and were then transferred to "high" temperature developed more rapidly than "controls" which had been at "high" temperature all along. Again this is just what would have been expected with a weak diapause. The "diapause" had disappeared from the "experimentals" while they had been at "low" tempera-ture, but not from the "controls" which had been at "high" temperature all along. Also thoroughly characteristic of a weak diapause is the gradient with temperature which is indicated by Figure 6.16, *B*. The higher the temperature at which the "experimentals" were placed, the more their advantage over the "controls." This is because some diapause-development would be possible at the lower temperatures in the "high" range but not at the higher ones (sec. 4.31).

b) The harmful influence of extreme temperature is discussed in section 6.3. At present we wish to illustrate the principle that when temperature fluctua-tions include temperatures outside the favorable range, it is possible for tem-porary exposure to the extremes to harm the animal so that its normal speed of development is not maintained when it is returned to a favorable temperature. A continuous exposure of 16 days at 5° C. was sufficient to kill about half the eggs of *Locusta migratoria;* all the eggs were dead after an exposure of 32 days at 5° C. It is clear that 5° C. is outside the limits of the tolerable range for eggs of *Locusta*. Moderately long (sublethal) exposures to 5° C. result in a slowing-down of the speed of development when the eggs are returned to a favorable temperature. The degree of retardation depends upon the duration of exposure to 5° C., the stage of development of the embryo, and the particular favorable high temperature at which development subsequently proceeds. This is indicated by the data in Table 6.05, which is condensed from the more extensive

TABLE 6.05*

INFLUENCE OF EXPOSURE TO 5° C. IN RETARDING SPEED OF DEVELOPMENT OF EGGS OF *Locusta* IN RELATION TO DURATION OF EXPOSURE, STAGE OF DEVELOPMENT OF EMBRYO, AND TEMPERATURE OF INCUBATION

Incubation-Temperature.....	27° C.		30° C.		33° C.		37° C.	
Days at 5° C...............	8	16	8	16	4	8	4	8
Per cent development:								
0.................	−0.07	. . .	0.63†	. . .	0.46†	0.69†	0.57†	. . .
25.................	.02	0.61†	− .12	0.66†	− .22†	− .06	.17†	0.35†
50.................	0.75†	1.48†	0.70†	1.48†	0.40†	0.86†	0.76†	0.97†

* Figures in the body of the table are differences in incubation periods (in days) obtained by subtracting controls from treatments. A dagger indicates significance at $P = 0.02$. After Ahmad (1936).

data given by Ahmad (1936). The retardation in the speed of development indicated in the table must be due to the harmful influence of continuous exposure to low temperature. The temperature (5° C.) and the duration of exposure (in some instances no more than 4 days) are both moderate, judged by the standards of temperate climates; but *Locusta* is an animal of tropical and subtropical regions, where temperatures of this order are unlikely except as part of the diurnal fluctuations, that is, they are not likely to be experienced for more than a few hours at a time. High temperatures outside the favorable range may have a similar harmful influence. For example, Ludwig and Cable (1933) found that when pupae of *Drosophila melanogaster* were exposed for 1 day to 33° C., they were unable to maintain their normal speed of development when returned to 25° C. The stages most affected were male pupae which were 1, 2, or 3 days old and female pupae 1 or 2 days old.

c) With many species, particularly among the holometabolous insects, different stages of the life-cycle live at different times of the year or in different sorts of places and experience quite different ranges of temperature. It would be surprising if these stages responded similarly to temperature; the differences are often so obvious that experimental demonstration is not necessary. But with species in which all stages of the life-cycle live together in a seemingly uniform place, it is necessary to look into the matter a little more closely. For example, with the rice weevil *Calandra*, all stages may be found all the year round living together in the same handful of grain. The egg stage of *Calandra* required, on the average, 10.4 days to complete its development at 18.2° C. and 3.6 days at 29.1° C. This indicated a proportional increase of 2.84 in the duration of the egg stage as the temperature decreased from 29.1° C to 18.2°. For the same pair of temperatures the proportional increases in the duration of the larval, prepupal, and pupal stages were, respectively, 4.78, 2.06, and 3.77. These are large and significant differences (Birch, 1945c). The same principle may be illustrated by measuring the rate of oxygen consumption of the different stages. With larvae of *Calandra* (12 days old at 29° C.) the oxygen consumption per insect per hour increased from 3.59 cu. mm. at 15° C. to 14.95 cu. mm. at 32° C., giving a proportional increase of 4.17. For adults (7 days old at 29° C.) the relative increase for the same pair of temperatures was 5.55. With larvae of *Rhizopertha*, another inhabitant of stored grain, the oxygen consumption per insect per hour was 1.41 cu. mm. at 22° C. and 4.84 cu. mm. at 38°, indicating a proportional increase of 3.43. With adults, the proportional increase for the same pair of temperatures was 2.95. With the *Drosophila melanogaster* the duration of the egg stage at 27.8° C. was 17.8 hours, and at 18.2° C., 41.4 hours, indicating a proportional increase of 2.34. The proportional increase in the duration of the larval stage for the same pair of temperatures was 2.62 (Powsner, 1935). These examples were quoted because they all come from experiments that were done with unusual care and precision. Many others could be given which point to the same conclusion.

Eclosion from the egg and transformation to pupa or adult are but easily perceived end-points in a process which consists of many more stages than these. It may be more difficult to demonstrate that the latter also respond differently to temperature, but there is no reason to suppose that they would differ in this respect from the longer, better-defined stages that have been studied. In section 6.231, in the discussion of the logistic curve, further evidence was advanced toward this conclusion.

If, then, it is true that the successive stages in development are likely to respond differently to temperature, it must be recognized that this introduces an error which cannot be avoided. But experience suggests that this error may be small, particularly if the stages that are considered together are kept as few as practicable (sec. 6.231, especially Figs. 6.09 and 6.14 and Table 6.03).

d) We shall now consider the possibility that healthy development may proceed during short or intermittent exposures to extreme temperatures which would be harmful or perhaps lethal if experienced continuously. This is the reverse of what was discussed in paragraph *b* above. It has been investigated by Birch (1942) for eggs of *Austroicetes cruciata* and by Ludwig and Cable (1933) for pupae of *Drosophila melanogaster*. Post-diapause eggs of *Austroicetes* will develop and hatch at constant temperatures between 16° and 33° C. But at 33° the harmful influence of continuous high temperature is manifest from the flattening of the temperature-development curve about this point. When eggs were incubated on alternate days at 33° and 19° C., it was found that the rate of development at 33° C. was greater than when the eggs were exposed to this temperature continuously. The pupae of *Drosophila* developed and the adults emerged at constant temperatures between 15° and 33° C. At 10° and at 34° C. no flies emerged, although at 10° C. many puparia contained fully formed flies which had failed to emerge; a few were also found in the puparia at 34° C. When pupae were placed on alternate days at 10° and 20° C., it was estimated that about 2.4 per cent of the total development had proceeded at 10° C. Similarly, it was shown that a little development was possible during intermittent exposures to 8° but not at 7° C. The design of these experiments does not completely preclude the possibility of these results being due to a diapause-effect, as in *Calliphora* and *Locusta*, but the alternative explanation seems more likely in this case.

e) If the relationship between temperature and speed of development were linear, then the speed of development of an animal in an environment where temperature is fluctuating within the limits of the favorable range could be accurately determined by linear interpolation on the temperature-development curve, using the arithmetic mean of the fluctuating temperatures. This is, of course, the procedure followed in the classical method of "temperature-summation" (Shelford, 1927). But we have already shown in section 6.231 that the relationship between temperature and speed of development is distinctly curvilinear (Fig. 6.09). It follows that the estimates got by linear interpolation lack precision and should not be used when precise estimates of the speed of development are required.

But it does not follow that fluctuating temperature within the favorable range is not equivalent to the corresponding constant temperature. On the contrary, the little experimental work that has been done in this field indicates that fluctuating temperature within the favorable range is closely equivalent to the corresponding constant temperature; and a precise estimate of speed of development can be made by a modified temperature-summation method which takes into account the curvature of the graph (sec. 6.234). For example, in one experiment the eggs of *Austroicetes* were exposed on alternate days to the following pairs of temperatures, 19° and 31° C., 23° and 27° C., and 23° and 31° C. In every case the speed of development was the same as that found at the corresponding constant temperature (Birch, 1942). Pupae of *Drosophila* were exposed on alternate days to the following pairs of temperature, all within the favorable range, 21° and 29° C., 23° and 28° C., 15° and 25° C. In no case was there any difference from results at the corresponding constant temperature (Ludwig and Cable, 1933).

These results are typical of others that might be quoted, and it seems that, provided that the diapause-effect is not influencing the results, short-term (e.g., daily) fluctuations in temperature within the favorable range may safely be considered to be equivalent to constant temperatures in this range. This is not to say that the average speed of development is the same as that at a constant temperature equal to the mean of the fluctuating temperatures. This would imply that the relationship between temperature and speed of development is linear, which it is not (sec. 6.231). When the fluctuations include extremes outside the favorable range, it is necessary to consider the following possibilities: (a) that there is a diapause-effect, (b) that short exposures to extreme temperature may impair the animal's competence to develop at a favorable temperature, and (c) that healthy development may be possible during short exposures to extreme temperatures that would be harmful if the exposure were prolonged. Provided that these restrictions on the scope of data from experiments at constant temperature are recognized, there is no reason why this should not continue to be the chief method used in the laboratory to study the influence of temperature on the speed of development.

6.233   SPEED OF DEVELOPMENT AS AN ADAPTATION

We saw in section 6.1 and Figure 6.01 that the limits of the favorable range of temperature may be correlated with the range of temperature characteristic of the places where the species usually lives. A similar adaptation may be recognized in relation to the influence of temperature on the speed of development. For example, Ide (1935) studied the distribution of mayflies (Ephemeroptera) in mountain streams in Canada. Quantitative records were kept at intervals throughout the year of the mayflies living in six different "stations" along one such stream. A "station" was a situation where the mayflies usually breed, selected for its similarity to all the other "stations" in all important

TABLE 6.06*

SEGREGATION OF SPECIES OF MAYFLIES IN A MOUNTAIN STREAM IN RELATION TO TEMPERATURE

"Station"†	Temperature (° C.)	No. of Species	No. of Species Not Found Higher Up	No. of Species Not Found Lower Down
1.................	9.0	7	..	..
2.................	16.3	15	8	1
3.................	19.5	16	2	..
4.................	21.5	22	8	1
5.................	20.5	21	2	1
6.................	24.0	29	6	..

* After Ide (1935).
† Station 1 (the coolest) is near the source, and station 6 (the warmest) is farthest from the source of the stream.

respects except temperature. The temperature was lowest near the source, and the water became warmer toward the lower reaches of the stream. For example, on one typical summer day the maximal temperature of the water at the six "stations" varied from 9° for the one nearest the source to 24° C. for the one farthest from the source (Table 6.06). There were more species living in the lower (warmer) situations than in the higher (colder) ones. For example, of the 15 species normally found at station No. 2, there were 8 species which were never found in station No. 1. Similarly, 8 of the 22 species from station No. 4 were never found in any situation colder than this. There was no evidence that cold of the order experienced at any of these six situations was directly harmful to any of the species.

Hence it must be some other aspect of low temperature that is limiting the distribution of the mayflies. Now, mayflies complete one full generation each year. The adults emerge between June and August, depending on the species. They live for only a few days. The rest of the year is spent in the egg or nymphal stages in the water. Ide's researches make it quite clear that the chief adaptation enabling a species to inhabit a colder situation is its ability to develop more rapidly at low temperatures and thus to complete its life-cycle within the year in these colder situations near the source of the stream. The same

TABLE 6.07*

RELATIONSHIP BETWEEN SPEED OF DEVELOPMENT OF EGGS OF FOUR SPECIES OF *Rana* AND TEMPERATURE OF WATER IN WHICH THEY USUALLY BREED

SPECIES OF *Rana*	NORTHERN LIMIT OF DISTRIBUTION	BREEDING SEASON	USUAL TEMPERATURE OF WATER AT BREEDING TIME (° C.)	HOURS REQUIRED TO DEVELOP FROM STAGE 3 TO STAGE 20			RATIO OF COLUMN *b* TO COLUMN *c*
				12° C. (*a*)	16° C. (*b*)	22° C. (*c*)	
sylvatica	67°	Late March	10°	205	115	59	1.95
pipiens	60°	Early April	12°	320	155	72	2.15
palustris	51°–55°	Mid-April	15°		170	80	2.12
clamitans	50°	June	25°		200	87	2.29

* Moore (1942).

principle may be important in relation to the northern distribution of species, because the species living at high altitudes near the source of the stream were largely the same as those found near sea-level in high latitudes.

The same principle has been demonstrated in Amphibia (Moore, 1939, 1940*a*, 1942). The morphogenesis of the amphibian embryo may be followed through a series of well-defined stages. These were described for the frog *Rana sylvatica* by Pollister and Moore (1937). Using the time required to develop from stage 3 to stage 20 as a criterion of the speed of development, Moore compared four species of the genus *Rana* (Table 6.07). He found a close relationship between the temperature of the water in which the species normally breeds and the speed of development of the eggs at constant temperature. The eggs of species from the colder places (e.g., farther north, or earlier in the season) developed more rapidly at low temperatures than did those of species from warmer places (farther south, or later in the season). On the other hand, at moderate temperatures the increase in speed of development with rising temperature was less for the northern species. The most southern species, *R. clamitans*, was able to develop from stage 3 to stage 20 in 45 hours at 33° C.—a temperature which was lethal to the other three species.

6.234 THE SPEED OF DEVELOPMENT IN NATURE IN RELATION TO TEMPERATURES RECORDED
IN THE FIELD

This matter has, for a long time, been of interest to "applied" biologists, particularly entomologists. It would be useful to be able, by means of meteorological records, to predict the date of the emergence of the spring brood of an insect pest or the duration of a generation. But the matter is not so simple as it may seem at first. The raw data from which we usually have to work consist of (a) reasonably precise information, derived from laboratory experiments, about the time required to complete the different stages of the life-cycle at a number of constant temperatures: (b) information, derived from observations in nature, about the behavior of the animal, particularly the sorts of places where it lives and the seasons of the year when the various stages of the life-cycle are present: and (c) records of temperature and other components of weather. For a limited investigation it may be possible to measure the temperature in the precise situations where the animal lives. But often this will be impracticable. For example, if it is desired to refer back a number of years, the standard records kept by the meteorological services will, almost certainly, be the only ones that are available.

The first step is to arrive at a satisfactory law relating temperature to the speed of development (sec. 6.231). Nearly all the early attempts to solve this problem centered around the hyperbola and its reciprocal, the straight line. Reibisch (1902) plotted the reciprocal of the time required for the development of fish eggs against temperature, drew a straight line through the observed points, and defined the point where this line cut the abscissae as the "threshold of development." Sanderson and Peairs used the same method to determine the "developmental zero" for 12 species of insects. Krogh (1914) showed that the "developmental zero" so determined was not necessarily a true threshold, since quite appreciable development might occur at temperatures well below this point. He also showed that the speed of development at higher temperatures near the upper limits of the favorable range was appreciably slower than might be expected from an extrapolation of the straight line.

But the hyperbola may be accepted as descriptive of the relationship between temperature and the time required for development within a limited range of temperature, without necessarily attributing any biological significance to a (see below). Simpson (1903) was probably one of the first to develop the concept of the "thermal constant" expressed in units of "day-degrees." This follows from the equation of the hyperbola, $K = y(x - a)$, where K is the thermal constant; y is time required to complete development; and, since a is the "developmental zero," $x - a$ is the "effective temperature." Since the reciprocal of the hyperbola is a straight line which cuts the abscissae at a, it also follows that when x is variable (as, for example, temperature in nature), K may still be calculated by summing $d(x - a)$, so long as the appropriate temperature x may be assigned to each period d. As a crude approximation, a day may be taken as the unit for d, and the mean of the daily maximal and minimal temperature may be substituted for x. This is the basis for the familiar practice of "temperature summation." This method was used in the well-known papers of Glenn (1922, 1931) on the codlin moth. He introduced a correction for the falling-off in speed of development at high temperature which had been demonstrated by Krogh, but he ignored the existence of a true threshold somewhere below the theoretical one, a. Shelford (1927), in his monumental work on the same insect, made allowance for departures from linearity at both ends of the medial range but not in the medial range itself. That is to say, the linear transformation of the hyperbola remained the chief basis of his extremely complicated computation of "developmental units." The "developmental unit" was introduced to replace the "day-degree." It made allowance for humidity as well as temperature. But the chief refinement associated with this method was due to the substitution of the hour as the unit for which temperature was measured, rather than the day. We shall show below that the use of the mean daily temperature may introduce substantial errors.

The use of the logistic equation in place of the hyperbola allows for the curvature of the temperature-development relationship throughout the whole range of temperature. The logistic equation formed the basis of a quantitative study of the rate of development of the eggs of the grasshopper *Austroicetes* in the field (Andrewartha, 1944a). This is an instructive example, because in it one meets a number of the difficulties that usually arise in this sort of study. (a) The eggs of the grasshopper are laid in the soil at a depth of about 1 inch. So they experience temperatures which are quite different from those registered in the meteorological screen. But it was necessary to use standard meteorological records because it was desired to analyze the records for 50 years back. (b) The only records available for the zone where the grasshoppers occurred were daily maximal and minimal air temperatures. (c) On numerous occasions the minimal daily temperature was lower than the threshold below which no appreciable amount of development occurs. On these occasions development was proceeding for only part of the day. So it was unrealistic and inaccurate to use the mean temperature for the full day to estimate the amount of development. (d) And, finally, there were no records of soil temperature in the area where the grasshopper occurred. But there was a station about 150 miles away for which records of temperature in the soil were available. So, first of all, the data for this station were analyzed as if the grasshopper eggs had been located in the soil alongside the thermometer, and then the results were interpreted in terms of the air temperatures recorded in the area where the grasshoppers occurred. The analysis is best described in four steps.

   a) *Measurement of "effective" temperature.*—No appreciable development goes on in the eggs of *Austroicetes cruciata* at temperatures below about 12° C. But the daily minimal temperature of the soil at a depth of 1 inch was frequently below this. So a horizontal line was drawn on the thermograph chart, cutting off that part of the graph exceeding this temperature each day. Development was considered to have been proceeding only during that part of the day which was so delimited. The best estimate of the mean temperature for this part of the day was got by measuring the area with a planimeter and divid-

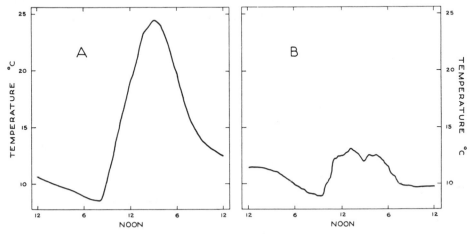

Fig. 6.17.—Showing thermograph records of the daily march of temperature in the soil at a depth of 1 inch at the Waite Institute, Adelaide. *A*, characteristic record for a bright sunny day in spring (September), and *B*, characteristic record for a dull cloudy day in winter (July).

ing this area by the length of the base line. Alternative estimates of the mean "effective" temperature were made (1) by taking the arithmetic mean of the maximum plus 12° C. and (2) by reading from the thermograph chart the maximal and minimal temperature for each 2-hourly period each day during the period when development was considered to be proceeding and taking the arithmetic mean for all these readings. The three different estimates are compared in Table 6.08. Both short-cut methods underestimated the mean "effective" temperature compared with the "true" mean temperature measured with the planimeter. But the second estimate (col. C of Table 6.08) came closer to the "true" mean temperature estimated with the planimeter than did the first. This is because the temperature did not change uniformly with time (Fig. 6.17). The first method takes much less account of the curvature and irregularities in the thermograph record.

b) Estimates of the amount of development.—The logistic curve was used to express the relationship between temperature and the speed of development (expressed as per cent of total development completed per day). The amount of development completed in a month was estimated by three different methods, and then, for purposes of comparison, this was expressed, in Table 6.09, as the mean daily speed of development for each month. (A) the unit of time was taken as 2 hours. The mean temperature for each 2-hourly period was calculated from the thermograph record. The speed of development appropriate to each 2-hourly period was read off directly from the temperature-development curve. These quantities were then averaged to give the mean speed of development for the day. (B) That part of the day during which development was considered to be proceeding was taken as the unit. The mean temperature was calculated as the maximum plus 12° C. divided by 2. The appropriate speed of development was read off the temperature-development curve. And then, since this

TABLE 6.08*

"Effective" Daily Temperature Estimated by Three Methods

	"Effective" Daily Temperature (° C.) 1 Inch under Soil				
Month	Planimeter (A)	Max. + 12 / 2 Daily (B)	Max. + Min. / 2 2-Hourly (C)	Difference A − B (D)	Difference A − C (E)
1938 June	14.42	14.09	14.24	+0.33	+0.18
July	15.15	14.60	15.00	+ .55	+ .15
Aug.	15.86	15.31	15.88	+ .55	− .02
Sept.	18.32	17.79	18.14	+ .53	+ .18
1939 June	15.42	14.89	15.19	+ .51	+ .21
July	14.59	14.21	14.50	+ .38	+ .09
Aug.	14.81	14.51	14.68	+ .30	+ .13
Sept.	17.69	17.02	17.52	+ .62	+ .17
Mean	. . . .	. . . .	. . . .	+0.48	+0.14

* After Andrewartha (1944a).

rate might apply to only part of the day, the appropriate correction was applied to express the result as the daily speed of development. (C) The day was the unit. The mean temperature was calculated by taking the mean of the maximum and the minimum. The mean daily speed of development was then read off directly from the temperature-development curve.

The results of the three methods are compared in Table 6.09. Method A, which, being based on 2-hourly intervals, makes full allowance for the threshold for development and also allows for most of the curvature in both the thermograph record and the temperature-development curve, leaves little room for error and may be used as a standard by which the other two methods may be judged. Method B, which allows for the threshold but ignores the curvature both in the thermograph record and in the temperature-development curve, underestimates the amount of development by amounts ranging from 5.7 to 24.0 per cent. Method C, which ignores the threshold and the curvature of both the temperature record and the logistic curve, was, as might have been expected, grossly misleading. It underestimated the amount of development by as much as 68 per cent in one month. Neither method seems sufficiently precise.

(c) *The correlation between maximal temperature and amount of development per day.*—Since the mean daily temperature, even when corrected for the threshold of development, was inadequate, it was necessary to carry the analysis one step further. The "true" daily speed of development (i.e., the daily figures from which the monthly means in col. A of Table 6.09 were derived) was correlated with the daily maximal temperature of the soil at a depth of 1 inch. A close curvilinear relationship was demonstrated (Fig. 6.18). This relationship would be quite consistent if the thermograph curve were the same shape each day. The variability was associated with .cloud, wind, and other meteorological events which modified the shape of the thermograph curve. The variability was, however, not great, and it was possible to use the daily

TABLE 6.09*

ESTIMATED MEAN DAILY RATE OF DEVELOPMENT OF EGGS OF *A. cruciata.*

MONTH	MEAN DAILY SPEED OF DEVELOPMENT AS PER CENT DEVELOPMENT PER DAY				
	Max. + Min. / 2 / 2-Hourly (A)	Max. + 12 / 2 / Daily (B)	Max. + Min. / 2 / Daily (C)	Difference as Per Cent of Column A	
				B − A	C − A
1938 June........	0.35	0.33	0.09	5.7	68.5
July.........	0.52	0.44	0.18	15.4	65.4
Aug.........	0.82	0.68	0.54	17.1	34.2
Sept.	2.22	1.97	2.14	11.3	3.6
1939 June........	0.60	0.54	0.51	10.0	15.0
July.........	0.44	0.39	0.18	11.4	59.1
Aug.........	0.50	0.38	0.26	24.0	48.0
Sept........	1.80	1.47	1.65	12.8	8.4

* Column A is based on mean temperature for 2-hourly period; column B on mean "effective" temperature; and column C on mean daily temperature. After Andrewartha (1944a).

maximal temperature of the soil at a depth of 1 inch to get a reasonably precise estimate of the daily speed of development.

d) The relationship between maximal temperatures of soil and air.—The calculations up to this point have been based on thermograph records of temperature in the soil at a station some 150 miles away from the zone where the grasshoppers occur. In this zone no records were kept of temperature in the soil. It was necessary to relate the temperature of the air to the temperature of the soil. This was done by comparing the two at the station where they were both kept, with the hope that the same differences would recur elsewhere in the same general climatic region. This was the step where there was the greatest room for error. It would not have been necessary, had there been adequate records of the temperature in the soil available for the area where the work was to be done. But this is the sort of difficulty which is all too familiar to ecologists.

The analysis which has just been described formed part of an ecological investigation aimed at understanding the reasons for the known distribution and abundance of the grasshopper *Austroicetes.* In this particular case the estimated date of hatching of the nymphs in the spring was calculated for each year for 50 years back. In the few years in which the "expected" date of

hatching could be compared with an observed date of hatching, the two estimates agreed within a few days.

The purpose of Shelford's (1927) study of the codlin moth was different. He
wanted to provide a method whereby farmers might be able to anticipate the
date of the emergence of the codlin moth each year, so that they might more

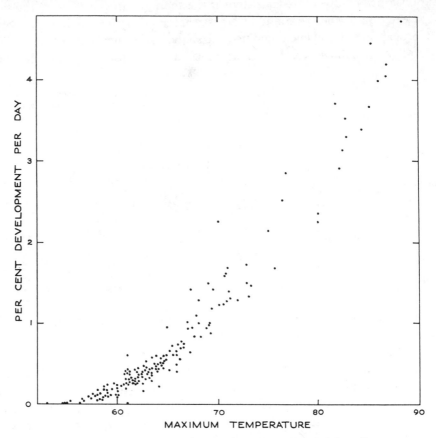

FIG. 6.18.—Showing the curvilinear relationship between maximal daily soil temperature at a
depth of 1 inch and the mean daily speed of development of eggs of *Austroicetes*. The maximal temperature was read from thermograph records. The speed of development was computed for each
2-hour period throughout the day and then averaged. (After Andrewartha, 1944*a*.)

accurately time their spraying program. This is an entirely different matter.
To be successful, the computations require to be highly precise. Sufficient has
been said in this section to indicate just how complicated and laborious the
measurements and calculations may need to be to get precise results from this
sort of analysis. It may well not be worth while as a routine practical measure:
in the case of the codlin moth, experience has taught that it is much better to
hang a few baited traps in the orchard and time the spraying program with
reference to the direct evidence provided by the appearance of moths in the

traps. But for a research project, as in the case of the work with *Austroicetes*, which has just been described, we may be more interested in the past than the future, and the labor of getting the most precise estimates possible may be well worth while. It is in this connection that the principles discussed in this section have their chief interest.

6.24 *Influence of Temperature on Fecundity and the Rate of Egg-Production*

The limits of the favorable range of temperature for egg-production and oviposition are often of the same order as those for the development of the immature stages, but need not necessarily be similar. Thus *Calandra oryzae*, living in wheat of 14 per cent moisture content, is able to complete its life-cycle at temperatures between 15° and 34° C. Oviposition also occurs over this range, but so few eggs are laid at 15° and 34° that it would be more realistic to consider this process to be limited to a rather narrower range, say 17°–33° C. (Birch, 1945c, d). With *Microbracon hebetor*, Harries (1937) found that the development of the immature stages could be completed between 12° and 32° C. but that oviposition was virtually restricted to the range 16°–36° C., although a few eggs were laid at 14° C.

The number of eggs laid during an animal's lifetime (i.e., its fecundity) may be strongly influenced by components in the environment other than temperature, notably moisture, food, and the number of other animals present in the environment. The daily number of eggs (i.e., rate of egg-production) depends upon these things, too, and, in addition, is influenced by the age of the animal. For example, Dick (1937) studied oviposition in a number of beetles and found that the fecundity and rate of egg-production were influenced differently by age for different species; in some species copulation caused an increase in both the fecundity and the rate of egg-production; food, humidity, and in some species the frequency of suitable situations for oviposition were also important. These considerations complicate the gathering of experimental data on the influence of temperature on fecundity and rate of egg-production and also the interpretation of field observations on this subject. Birch (1945d) studied the oviposition of two beetles commonly found in stored grain, *Calandra oryzae* and *Rhizopertha dominica*, in a carefully controlled environment and was able to measure not only the direct influence of temperature but also its interaction with the moisture content of the food, the number of other animals in the environment, and the age of the ovipositing females (secs. 3.12, 3.4, and 9.222). The influence of temperature in relation to the number of other animals of the same sort may be seen in the information set out in Table 6.10. The number of eggs laid reached a maximum at a median temperature of 26° C. for the sparser population and 29° for the denser one. At temperatures above and below this, fewer eggs were produced. With *Thrips imaginis*, very few eggs were laid at 8°, but within the range 13°–23° C. there was a tendency for more eggs to be laid, the higher the temperature. Owing to the variability of the data, these differences were nonsignificant, but it is likely that with larger samples the trend could be established more precisely (Andrewartha, 1935). Similarly with the European cornborer *Pyrausta nubilalis*, Vance (1949) found that the moths laid more eggs at 29° than at 21° or 32° C. (Table 6.11).

TABLE 6.10*

TOTAL NUMBER OF EGGS LAID BY *Calandra oryzae* IN WHEAT
OF 14 PER CENT MOISTURE CONTENT AT DIFFERENT
TEMPERATURES

Temperature (° C.)	1 Insect per 10 Grains	1 Insect per 50 Grains
23	264	266
26	265	384
29	296	344
32	197	. . .
35	5	. . .

* After Birch (1945*d*).

The aphid *Toxoptera graminum* is ovoviviparous. Young are produced throughout adult life, but rather more rapidly by the young adult than by the older one. The total number produced during the lifetime of one female depended on the temperature at which it was living (Wadley, 1931). Figure 6.19 illustrates this relationship. With both the winged and the wingless form of

TABLE 6.11*

INFLUENCE OF TEMPERATURE ON NUMBER OF
EGGS LAID BY *Pyrausta* KEPT AT 96 PER CENT
RELATIVE HUMIDITY AND OFFERED WATER TO
DRINK

Temperature (° C.)	No. Fertile Eggs per Moth
21	708
25	758
29	823
32	533

* After Vance (1949).

Toxoptera, the shape of the curve relating temperature to fecundity is the same, but that for the winged form is distributed over a lower range of temperature than that for the wingless form.

These four examples are typical of others that could be quoted. They indicate that the fecundity of insects tends toward a maximum at a moderately high temperature, which may be relatively near the upper limit of the favorable range. As the upper or lower limit of the favorable range is approached, the fecundity declines abruptly (Fig. 6.19).

The influence of temperature on the rate of egg-production by insects (as distinct from the total eggs produced during a lifetime) has been reviewed by Harries (1939). The seven examples which he used consistently indicated the same trend. Below a certain well-defined lower limit the rate of egg-production was virtually zero. As temperature increased, the rate of egg-production increased up to a maximum and then decreased at still higher temperatures, until the upper limiting temperature was reached, when the rate again became zero. For the temperature range within which the rate of egg-production was increasing, the trend with temperature followed a sigmoid course (Fig. 6.20). Norris (1933) found that most females of *Ephestia kühniella*, which had been mated with one male only, produced fertile eggs at 27° and at 30° C., provided that the prepupae and pupae of both males and females had been reared at a

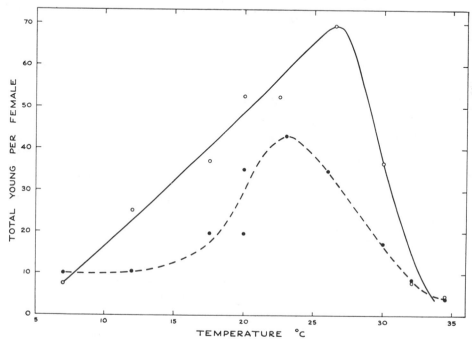

Fig. 6.19.—Showing the influence of temperature on the fecundity of the aphid *Toxoptera graminum*. *Complete line,* the winged form; *broken line,* the wingless form. (Data from Wadley, 1931.)

favorable temperature lower than 27° C. As the temperature at which the pre-pupae and pupae were reared increased from 27° to 30° C., the proportion of females which laid no fertile egg increased from 10–59 per cent to 96–100 per cent. There was a certain variation between "cultures," but these experiments showed that exposure to unfavorably high temperature during the prepupal and pupal stages made the adults sterile.

The results from experiments with constant temperatures may not necessarily be equivalent to what happens with fluctuating temperatures in nature. There is evidence that, at least in some circumstances, a short period at low temperature may result in an increased rate of egg-production on return to a higher temperature (Dick, 1937). Figure 6.21 illustrates this for *Tribolium confusum*. After 100 days at a constant temperature of 27° C., the rate of egg-production had fallen from an initial rate of more than 8 per day to about 3 per day. On the 101st day the insects were placed at 18° C. and kept at this temperature for 8 days. On being returned to 27° C. on the 109th day, the insects began producing eggs more rapidly, quickly reaching a peak of about 5 eggs per day, which was nearly double the rate for the period just before they had been placed at low temperature. At a constant temperature of 27° C. the rate began to decline again but required about 70 days to decrease to the origi-

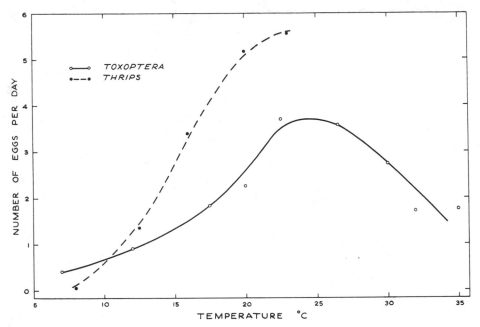

FIG. 6.20—Showing the relationship between temperature and speed of egg-production for *Thrips imaginis* and *Toxoptera graminum*. (Data from Wadley, 1931; and Andrewartha, 1935.)

nal figure of 3 eggs per day. There can be no doubt that in this case a short exposure to 18° after the insects had been living at 27° for 100 days caused a marked increase in the rate of egg-production at 27° and that this increase was sustained for at least 60 days.

During a 10-year study of the European cornborer *Pyrausta nubilalis*, in the field in Ontario, Stirret (1938) accumulated a great volume of quantitative data on the flight activity, fecundity, and oviposition-rate of this species. The results of this study of a natural population confirmed the results of the laboratory experiments discussed in this section. The fecundity was estimated by counting the numbers of eggs laid each night on a sample plot of 100 plants of corn, and at the same time the number of moths flying in the plot was estimated independently. The average number of eggs per moth varied from 2.5 in 1928 to 15.2 in 1930. There was an apparent relationship with temperature, which suggested that most eggs were laid in those years when the mean daily air temperature during the flight period was about 23° C. and the atmospheric humidity was equivalent to a saturation deficiency of about 7 mm. of mercury.

The males of some mammals become sterile when they are exposed continuously to high temperature. Gunn, Sanders, and Granger (1942) found that in certain hot districts in western New South Wales rams were sterile during the hot season. The semen began to degenerate about 4 weeks after the date

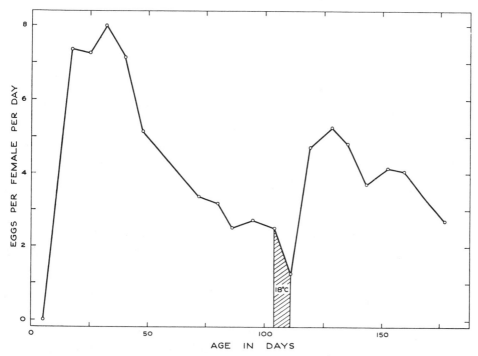

FIG. 6.21.—Showing mean daily rate of egg-production of *Tribolium confusum* for 180 consecutive days. The insects were at 27° C. for all the time except for 8 days from the 101st to the 108th day. Note sharp and sustained increase in rate of egg-production after 109th day. (Data from Dick, 1937.)

when the daily maximal temperature began consistently to exceed 32° C. Testes which had degenerated became normally active again about 2 months after the hot season had ended. By keeping rams in hot rooms and by insulating the testes during the cool season, Gunn, Sanders, and Granger showed that temperature was the chief cause of the sterile condition. Moore and Quick (1924) and Cowles (1945) got similar results with other species of mammals by heating the testes artificially or by transplanting them into the body. Cowles (1945) suggested that the habit of mating in the spring, which is commonly found in birds and other warm-blooded animals which live in the temperate zone, might be an adaptation which is associated with the tendency for the males to become sterile during hot weather.

6.3   THE LETHAL INFLUENCE OF TEMPERATURE OUTSIDE THE FAVORABLE RANGE

In the sense in which we have so far been using it, the favorable zone is delimited by the temperatures above and below which growth and multiplication cease to be possible, even though the animal may live for a long time at these temperatures. But the favorable zone may be considered in another way, in which the emphasis is placed on the survival of the *individuals* rather than on the continuance of the *population*. In sections 6.31 and 6.32 we discuss certain experiments in which animals were exposed to lethal low or lethal high temperatures.

6.31   *The Lethal Influence of Low Temperature*

The lethal influence of low temperature has been the subject of much study, especially by entomologists in Europe and North America; for the survival-rate during winter, together with the multiplication during summer, may largely determine the numbers of a certain pest in a certain area. Species from temperate climates often have a stage in the life-cycle which is especially adapted to survive during exposure to extreme cold: and these are said to be "cold-hardy." On the other hand, with species from tropical or subtropical climates, all stages in the life-cycle and, with species from temperate climates, the stages of the life-cycle that are usually present during summer may lack the capacity for becoming dormant when the temperature falls below that which favors active development. These forms are, as a rule, readily killed by exposure to moderate temperature, often well above 0° C.

6.311   THE THEORY OF "QUANTITY FACTOR" AND "INTENSITY FACTOR"

Because the animal may have to be exposed to moderately low temperature for a considerable period before it dies, ecologists sometimes speak of the "quantity factor" as distinct from the "intensity factor," which refers to the instantaneous killing of a cold-hardy individual by the freezing of its tissues (Payne, 1926a, b, 1927a). Payne coined these terms after studying the lethal influence of cold on representative insects from three groups which in nature are exposed to very different levels of cold. The oakborers were taken to represent those which are exposed during winter to temperatures many degrees below 0° C.; certain aquatic forms were taken to represent those that might experience temperatures as low as 0° C. but not below; and certain insects, usually considered to be of subtropical origin, which commonly infest stored grain, served as examples of the group which is used to living in situations where the temperature is usually well above 0° C. The "quantity factor" refers chiefly to the lethal influence of "low" temperatures (usually above 0° C.) on this third group: the term implies that the exposure to "cold" requires to be maintained for a considerable period—days or weeks. It is well known that representatives of the first two groups may survive unharmed for many months at these temperatures. On the other hand, most of them die instantly when their tissues freeze, as, of course, do the members of the third group also. The temperature required to bring about this freezing of the tissues is the measure of the "intensity factor" (see below, next section).

This theory seems to have dominated the outlook of most students of this subject up to the present time. See, for example, the summaries published by Uvarov (1931), Salt (1936), Mellanby (1939), Wigglesworth (1939, p. 364), and Luyet and Gehenio (1940). But, as Salt (1936, 1950) pointed out, the division of insects into those which respond to the "quantity factor" and those which respond to the "intensity factor," which is implicit not only in the original theory but also in most of the later work based on it, is unreal. Representatives of the "nonresistant" group (e.g., *Ephestia*) die after very short (virtually instantaneous, e.g., 10-minute) exposures to nonfreezing temperatures below 0° C. On the other hand, representatives of the "resistant" group may also succumb to exposure to temperatures that do not result in the freezing of their tissues—provided that the temperature is low enough and the exposure long enough. Moreover, the temperature required to produce freezing depends upon the duration of exposure; it is not, as has hitherto been widely assumed, independent of time. As a consequence, it seems that the conventional "under-cooling point," got by rapidly cooling the animal until a "rebound" is observed (see below), indicates only the temperature required to bring about freezing after an instantaneous exposure; it gives little information about the time-temperature responses of the animal to cold. It is clear that the distinction between "quantity factor" and "intensity factor" is not so simple as it has seemed.

The consequences for ecology of the long predominance of this too simple hypothesis is that the study of the lethal influence of nonfreezing temperatures has unfortunately been largely restricted to temperatures above 0° C. And the information gathered about the lethal influence of freezing temperatures has been unnecessarily abstract, dealing with the temperature required to produce instantaneous freezing, but ignoring for the most part the interaction between temperature and time which is an important feature of nature. It is necessary to consider the following principles: (*a*) Some sorts of animals may continue to live even after their tissues have been frozen; nevertheless, they succumb to adequately prolonged exposure to adequately low temperature. (*b*) All other sorts die instantly when their tissues freeze, the time and temperature required to produce freezing varying widely. (*c*) Many, perhaps most, sorts can be killed by adequate exposure to "low" temperatures which are not low enough to freeze the tissues; but the low temperature above which the animal may live indefinitely varies enormously; it may be well above 0° C. for some and well below 0° for others. (*d*) In all three categories the death-rate depends upon both time and temperature. In the sections which follow we give a few selected examples to illustrate these principles.

6.312   THE LETHAL INFLUENCE OF NONFREEZING TEMPERATURES

It may reasonably be supposed that for any particular individual there is a precise limit to the time during which it may survive exposure to a particular lethal temperature. If we could observe this limit directly for a sufficiently large number of individuals drawn at random from the population which we choose to study, we would be in a position to calculate directly, in the ordinary way, the mean duration of the fatal exposure and ascribe a variance to it. These two statistics would adequately describe the lethal influence of temperature on the population under study. But unfortunately, especially with high temperature and nonfreezing lethal low temperature, we usually have no way of telling by direct observation the precise moment when death has been irrevocably determined. It may be necessary to wait for a day or two, sometimes much longer, after the exposure to lethal temperature has been terminated and the animals returned to a favorable temperature, in order to see which ones will die (sec. 6.323). In this case the statistical method known as "probit analysis" is appropriate.

If the problem is to measure the duration of exposure to a certain temperature which is required to kill any proportion of the population, we begin by exposing a number, say six samples of 50 or more animals each, to this temperature for varying periods. The periods are chosen with the expectation that a small proportion of the animals will die as a result of the shortest exposure and that the longest exposure will kill nearly all those experiencing it. Alternatively, the problem may be to measure the particular temperature required to kill any proportion of the animals when they are exposed to it for a certain period. Then the procedure is varied to the extent that all the samples are exposed for the same period but to a range of temperatures chosen with the expectation that few will die as a result of exposure to the highest temperature and that exposure to the lowest temperature will kill most of them. In either case the animals are then removed to a favorable temperature. After a suitable interval, say several days or longer, the number of deaths in each sample is recorded.

If these numbers, expressed as proportions, are plotted against the treatment (or dosage), they will fall along a sigmoid curve, which is the familiar "dosage-mortality" curve of the toxicologist. This curve is the one that can also be obtained by progressively summing the ordinates of the ordinary frequency polygon. Provided that the frequency-distribution of deaths against "dosage" is normal or nearly so, the sigmoid "dosage-mortality" curve becomes a straight line when the proportions are transformed to probits (Finney, 1947).

If the distribution is not normal, it may usually be made so by transforming the dosage to logarithms or some other appropriate scale.

Once the straight line has been formed, it is easy to read off, by linear interpolation, the dosage (temperature or duration of exposure) which would be expected to kill any proportion of the population. Usually the most instructive figure is the particular dosage required to kill 50 per cent of the population. Strictly, this measures the median of the distribution, but it is also a good estimate of the mean. This is equivalent to the mean we would calculate if we could write down by direct inspection of each individual the duration of exposure just required to kill it. We may sometimes wish to know the dosage required to kill some other proportion of the population. There is no difficulty in this, but it must be remembered that, unless the size of the samples is increased, the estimates will be less precise. To estimate the dosage required to kill 95 per cent the samples must be 3 times, and for 99 per cent 10 times, as large as those required to estimate, with the same degree of accuracy, the dosage required to kill 50 per cent.

The "probit" analysis has become standard practice in "applied" laboratories where insecticides or drugs are being tested. Unfortunately, in the field of "fundamental" ecology, where it would often be equally appropriate, the method has hardly been used, presumably because far too many biologists have been unfamiliar with the statistical procedures which enable precise methods to be used in this subject. The methodological mistakes which recur most frequently in the literature are: (*a*) the animals have been kept at the unfavorable temperature until they are visibly "dead," notwithstanding the fact that death must have been determined long before it could be detected in this way; (*b*) we are told the duration of the exposure required to kill all the animals in the sample. This depends on the size of the sample and is not a statistic that can be measured precisely by any method of sampling short of including the whole population in the sample. The probit method enables the investigator to extract most information from his results. Failure to design experiments so that the method can be applied leaves the reader groping for the little bit of information that can be got. It is very little extra trouble to design suitable experiments, and the subsequent analysis is not difficult. It is fully set out in Finney (1947).

The bedbug *Cimex* may be taken as an example of a species which is not capable of becoming dormant at temperatures below the favorable range for development. The influence of nonfreezing temperatures down to 1° C. on the survival-rate of eggs of *Cimex* has been discussed in detail by Johnson (1940). This is a paper which should be studied in the original by the student who is interested in methodology. Using the probit transformation, Johnson was able to make exact comparisons between the survival-rates of eggs of different ages and histories, exposed to various low temperatures at several different levels of humidity.

In one series of experiments, eggs of a uniform age (0–24 hours at 23° C.) were exposed to constant temperatures between 1° and 12° C. for periods ranging from 7 to 50 days. At the conclusion of each exposure the eggs were removed from the low temperature and incubated at 23° C. Failure to hatch was taken as the criterion for death. A typical experiment is set out in Table 6.12, which shows the results when eggs were exposed in a moist atmosphere (saturation deficit about 1 mm. of Hg) at 4.2°, 9.8°, and 11.7° C. The mortality is expressed both as a per cent and as a probit. The former is transformed to the latter simply by reference to the appropriate table (in Finney, 1947) or Fisher and Yates (1948). Table 6.12 also includes the estimated values of b, L.D. 50, and L.D. 99.99. The first is the coefficient for linear regression of probit on exposure time, i.e., it measures the slope of the probit line. The second and third are the estimated durations of exposure corresponding to death-rates of 50 and 99.99 per cent, respectively. The variances for b and L.D. 50 are also included because these quantities are necessary in the calculation of the statistical significance of the results.

TABLE 6.12*

DEATH-RATE OF EGGS OF *Cimex* EXPOSED FOR VARIOUS PERIODS TO 4.2°, 9.8°, AND 11.7° C. IN MOIST AIR

	4.2° C.			9.8° C.			11.7° C.	
Exposure (Days)	Death-Rate		Exposure (Days)	Death-Rate		Exposure (Days)	Death-Rate	
	Per Cent	Probit		Per Cent	Probit		Per Cent	Probit
9.........	4.2	3.27	7........	10.1	3.72	7........	9.4	3.7
16.........	16.9	4.04	14........	7.5	3.56	14........	9.7	3.7
23.........	52.5	5.06	21........	19.9	4.15	21........	36.7	4.7
30.........	86.3	6.09	28........	65.8	5.41	29........	59.2	5.2
35.........	94.5	6.60	35........	86.3	6.09	35........	69.4	5.5
40.........	91.8	6.39	42........	100.0	8.72	42........	90.7	6.3
44.........	94.5	6.60				46........	94.0	6.6
50.........	100.0	8.72				49........	96.2	6.8
b...........	0.1139 (0.000151)†			0.1273 (0.000241)			0.0800 (0.000058)	
L.D. 50......	22.9	(1.120)		25.8	(0.774)		26.7	(1.409)
L.D. 99.99	55.6			55.0			73.3	

* After Johnson (1940).
† Figures in parentheses are variances.

Figure 6.22, A, shows the death-rate, in per cent, plotted against duration of exposure, and Figure 6.22, B, shows the same data plotted as probits. The advantages of transforming the data to the scale in which the regression becomes linear are that it enables a precise estimate of the exposure time required to kill any specified proportion of the eggs; and it makes it possible to attribute a significance to the observed differences. Figure 6.22, B, indicates that the probit lines for 4.2° and 9.8° C. are nearly parallel and that both slope more steeply than the probit line for 11.7° C. On the other hand, if we compare the co-ordinates on the abscissae for probit 5 (ie., the value for L.D. 50), we find that 9.8° and 11.7° C. come close together, whereas 4.2° is well away to the left. These two statistics, namely, L.D. 50 and *b*, are the two most instructive ones to extract from the probit line. The former is the most reliable

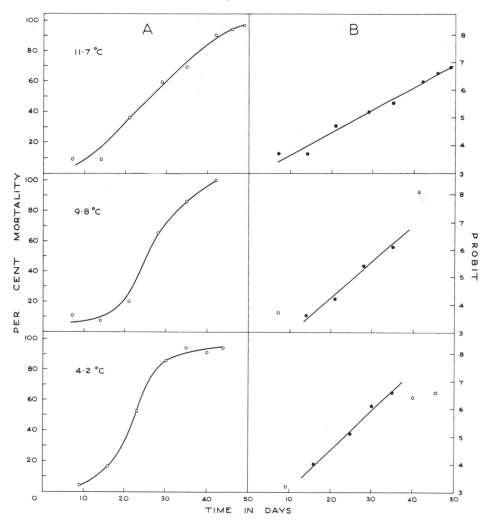

Fig. 6.22.—Showing the death-rate for eggs of *Cimex* exposed to 4.2°, 9.8°, or 11.7° C. for a varying number of days. *A*, the per cent mortality is plotted against time, giving a sigmoid curve, *B*, by transforming per cent to probit, the sigmoid curve becomes a straight line. (Data from Johnson, 1940.)

measure of the animals' capacity to survive at a given temperature, and the latter is the best comprehensive measure of the relative toxicity of short and long exposures. A large value for *b*, i.e., a steep slope to the probit line, indicates that there is relatively little "scatter" between the exposures that kill few and many; conversely, a small value for *b*, i.e., a gently sloping probit line, indicates a wide margin between the exposures required to produce these extremes in the death-rate.

It will be instructive to see what measure of significance may be attributed to the differences which we have pointed out above. Table 6.13 sets out the appropriate statistics. The difference between the regression coefficients for 4.2° and 9.8° C. is less than its standard deviation; hence, clearly, we may consider these lines to be parallel. But the difference between the regression co-

TABLE 6.13
SIGNIFICANCES OF DIFFERENCES IN VALUES OF b AND L.D. 50 GIVEN IN TABLE 6.12

Comparisons	Differences	S.D.	t	P
b for 9.8° C. − b for 4.2° C..................	0.0164	0.098	..	...
b for 11.7° C. − b for 4.2° C..................	0.0339	0.0145	2.3	<0.05
L.D. 50 for 11.7° C. − L.D. 50 for 4.2° C........	3.8	1.590	2.4	< .05
L.D. 50 for 9.8° C. − L.D. 50 for 4.2° C........	2.9	1.376	2.1	<0.1>0.05

efficients for 11.7° and 4.2° C. exceeds its standard deviation 2.3 times. Such a difference would have occurred by chance in less than 5 per cent of tests; hence we may consider that these lines differ in slope. This means that the range in the extreme exposure times which, on the one hand, just allowed all the eggs to survive and, on the other hand, was just sufficient to kill them all was the same at 4.2° as at 9.8° C., but was narrower at both these temperatures than at 11.7°. The wider range at 11.7° may be associated with the fact that some development is possible at 11.7° (although the eggs cannot hatch at this temperature) and the fact that the survival-rate depends upon the stage of development of the eggs (see below).

The difference between L.D. 50 for 11.7° and 4.2° C. exceeds its standard deviation in the ratio 1:2.4; with 12 degrees of freedom, this gives a probability of something less than 5 per cent that this difference has occurred by chance, and it may reasonably be concluded that eggs die more rapidly at 4.2° than at 11.7°. The difference between L.D. 50 for 9.8° and 4.2° exceeds its standard deviation in the ratio 1:2.1; a difference of this magnitude might have occurred by chance about 7 times in a hundred trials; and if we had to rely on this evidence alone, we should be cautious about accepting this difference as proof that 9.8° C. is more lethal than 4.2° C. In fact, Figure 6.23 shows that the curve is relatively steep between these temperatures, indicating that the death-rate is changing rather rapidly with respect to temperature.

The variances for L.D. 99.99 (i.e., the estimated exposure required to kill all the eggs) are not given in Table 6.12. They would inevitably be much larger than those for L.D. 50, and it may well be that none of the values given in Table 6.12 for L.D. 99.99 would differ significantly from any other. Although the L.D. 50 (i.e., the exposure required to kill half the animals in the sample) is the statistic that can be estimated most reliably; nevertheless, the estimate of L.D. 99.99 derived from the probit line is better than any that can be obtained by attempting to measure empirically the exposure required to kill all the animals. The chief reason for this is that in the empirical determination the exposure depends not only on the temperature but also on the size of the sample.

When the temperature was reduced below 4° C., a much shorter exposure was fatal. The time required to kill 50 per cent was shortest at the lowest temperature tested (namely 1.1° C.); it increased rather steeply up to 7° C.;

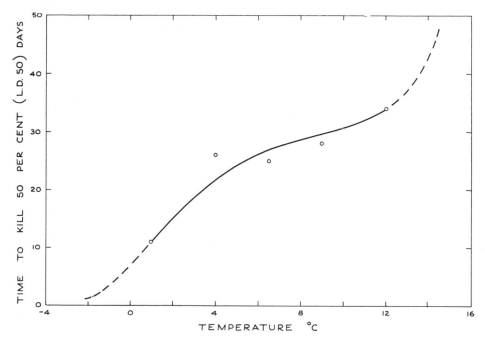

FIG. 6.23.—Showing the relationship between temperature and the time of exposure in days required to kill half the individuals in a sample of eggs of *Cimex*. The curve obviously approaches infinity at the lower incipient lethal temperature and approaches zero at some low temperature which was not measured. The broken lines are extrapolations which illustrate the limits toward which the curve is tending. (Data from Johnson, 1940.)

above 7° the curve flattened somewhat (Fig. 6.23). Since the eggs can hatch with less than 50 per cent mortality at temperatures above 15° C., this may reasonably be taken to be about the incipient lethal low temperature (i.e., the temperature above which the animal may live indefinitely without being killed by temperature); and in extrapolating the curve in Figure 6.23 we have drawn it asymptotic to 15° C. Johnson did not use temperatures lower than 1.1° C., so that the extrapolation of the curve at its lower end is largely hypothetical, but it is based on empirical data for other species.

The tissues of most insects probably have a freezing point in the range −1° to −2° C. If a batch of animals is cooled below the freezing point of the tissues, some will freeze and some will not; this is related to their capacity for "undercooling," which we discuss below. If they are of the sort which cannot withstand freezing, all those that freeze will die. But some of those which have not been frozen will have been fatally injured if the exposure to low temperature has been long enough. There are indications from the literature (e.g., Nagel and Shepard, 1934; Salt, 1936) that animals which lack the capacity to become dormant may be killed without being frozen by as little as 10–20 minutes' exposure to temperatures of the order of −10° C. For example, Salt (1936) showed that an exposure of 15 minutes to −15° C. could be fatal to adults of the beetle *Tribolium*, even though their tissues had not been frozen. About 75 per cent of the beetles died as a result of this exposure; of these, about 55 per cent had been frozen, leaving about 20 per cent that had definitely been killed without being frozen. To this figure must be added a proportion, which cannot be determined, of the 55 per cent which had accumulated a lethal dosage of cold before freezing. We do not know how many of these there might be, but it would certainly seem safe to count more than 20 per cent, perhaps 30 per cent, mortality due to the lethal influence of nonfreezing temperature during an exposure of 15 minutes to −15° C. Nagel and Shepard (1934) estimated that about half the eggs of *Tribolium* died from an exposure of 45 minutes to −6° C.; with sixth-instar larvae, half died after 30 minutes at −12° C. and 12 hours at −4° C. These figures overestimate the number of deaths due to nonfreezing temperatures, because at these temperatures some of the insects must have undercooled and frozen, and all of these may not have died without being frozen.

It seems reasonably certain that the L.D. 50 (due to nonfreezing temperature) continues to decrease with decreasing temperature and, at some low temperature, approaches zero. We have therefore extrapolated the lower part of the curve in Figure 6.23 on these grounds and have made it approach closely to zero. There need be little doubt that a generalized curve of this general shape describes the relationship between low temperature and the death-rate due to lethal effects other than freezing for the large class of animals which lack the capacity for dormancy when exposed to temperatures below the favorable range for development. The curve may vary in position and in other details for each particular sort of animal; also for different stages of the life-cycle of the same animal; or for the same stage which has been differently acclimatized or has in some other way had a different history.

Thus Johnson (1940) found that the survival-rate of the eggs of *Cimex* exposed to low temperature depended on the age of the egg. When eggs were exposed to 7.7° C. in moist air, there was little difference between eggs that were from 1 to 4 days old at 23° C. But after the fourth day they became more susceptible to cold as they became older (Fig. 6.24).

Johnson was unable to demonstrate any results from acclimatization. Eggs which had been produced and laid at 15°, 18°, and 23° C. were exposed to 10°

in moist air; the time required to kill 50 per cent was the same for all three lots. Robinson (1928) tried to acclimatize adults of *Calandra* by exposing them to a series of descending temperatures: 72 hours at 10°, 65 hours at 7.2°, 48 hours at 4.4°, and 36 hours at 1.6° C. These beetles were then exposed to −1.1° alongside controls which came to the low temperature directly from 23°. The time required to kill about 50 per cent of the controls was 90 hours, compared to 28 hours for those that had been "acclimatized." This result might have been

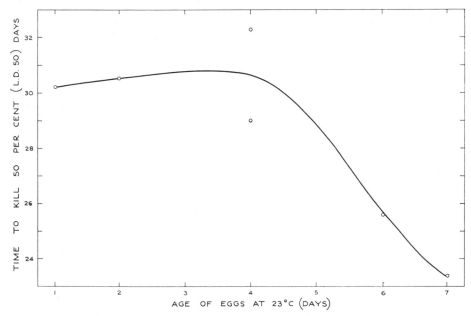

Fig. 6.24.—Showing the way in which the stage of the developing embryo influenced the survival time for eggs of *Cimex* held at 7.7° C. and 90 per cent relative humidity. (Data from Johnson, 1940.)

anticipated, for the "acclimatization" temperatures were below the incipient lethal low temperature for *Calandra*. During their so-called "acclimatization" the beetles were actually experiencing lethal temperatures and were accumulating a sublethal dose of cold; they died in a shorter time at −1.1° C. because they had already been "part-killed." This experiment is often quoted as proof that *Calandra* (and perhaps other similar animals) cannot be acclimatized. The temperatures at which it would be reasonable to acclimatize *Calandra* and other animals lacking the capacity for dormancy are those which lie just inside the favorable zone for development. So far as we know, this experiment has not been done, and we can not yet say whether this sort of animal does or does not possess the capacity to become acclimatized.

Most of the common insect pests of stored products are usually considered to have originated in tropical or subtropical zones. All those that have been studied appear to lack the capacity to become dormant in any stage of the life-cycle; they are therefore unable to survive exposure to temperatures below the range that is favorable for development. For example, Nagel and Shepard (1934) estimated the time required to kill 50 per cent of several stages of *Tribolium.* Their results are summarized in Table 6.14.

Robinson (1928) found that about half the sample of adult beetles of *Calandra oryzae* died after an exposure of 8 days at 7.2°, 4 days at 1.6°, or 2 days at −1.1° C. With adult beetles of *Oryzaephilus surinamensis*, 30 days at 10° C. or 4½ days at 2° were required to kill about half of those in the samples (Thomas and Shepard, 1940). Similar results have been reported for species that normally live in warm climates or warm situations and therefore are not usually exposed to the hazards of a cold winter. The fruit fly *Ceratitis capitata* is usually considered to have originated in tropical Africa. It has since spread to a number of warm, temperate areas, including South Africa, Australia, and the Mediterranean region of Europe; but it is not known from any cool, temperate region. Nel (1936) found that neither eggs, larvae, nor pupae could

TABLE 6.14*
ESTIMATED L.D. 50, IN HOURS, FOR DIFFERENT STAGES OF *Tribolium*

TEMPERATURE (° C.)	EGG			LARVA		PUPA
	1–24 Hours	48–72 Hours	120–32 Hours	3d Instar	6th Instar	
12............	118	258	157	...	...	...
7............	43	214	110	134	149	258
0............	5	....	...	...	...	...
−4............	1.5	3.5	...	9	12	...

* Ages expressed in terms of hours at 27° C. From Nagel and Shepard (1934).

survive exposure to temperatures in the range 0°–3° C. for more than 1–3 weeks. A significant proportion of the eggs of *Locusta migratoria* and pupae of *Muscina stabulans* were killed during exposures of 8–16 days at 5° C., but the same exposure did not reduce the survival-rate of pupae of *Calliphora erythrocephala.* The first is a subtropical species, the second ranges into cool, temperate zones, but the pupae are usually found in the warmer situations; the last is widely distributed in cool, temperate climates, and the pupae may occur in relatively exposed situations.

Many of the forms that live in temperate climates differ markedly in their resistance to cold at different stages of the life-cycle. The stage that is normally present during the summer may resemble the subtropical form that we have been discussing, in that dormancy is not possible and the incipient lethal low temperature corresponds more or less to the lower limit of the range favorable for development. The stage that is normally present during winter may be much harder to kill with cold, because it is able to become dormant when the temperature falls below that at which development is possible. For example, the Japanese beetle *Popillia japonica* overwinters in the second or third larval instar. In these stages it is capable of becoming dormant and may survive moderately low temperatures almost indefinitely. But both the adult and the first-instar larva lack this capacity. The newly emerged first-instar larva cannot survive 15 days at 10° C. (Ludwig, 1928). The small parasitic wasp *Diadromus collaris* usually overwinters as an adult, in which stage the females, in particular, are able to withstand several months at 4.5° C. But the larvae and pupae, which are the stages normally present during summer, die after 3 or 4 weeks at this temperature (Given, 1944). The red scale of citrus, *Aonidiella aurantii*, may overwinter in all stages in the milder parts of its distribution, but nearer its northern limits the adult tends to be the only stage to survive the winter. Munger (1948) placed all stages of the scale, including the very early "white-cap" stage, the "first moult," "second moult," gray adult, and mature adult in an outdoor "lath house" in California at the beginning of winter. All stages had been reared at 25° C. During the winter the temperature of the air in the lath house ranged from 0.5° to 7° C. By the end of winter, deaths among the adults amounted to 12 per cent; the deaths among other stages exceeded 50 per cent; from 73 to 96 per cent of the youngest ("white-cap") stage had died.

With insects, the capacity for dormancy seems to be most highly developed among those that are also able to enter the state of diapause (chap. 4). Diapause is often associated with the stage of the life-cycle that is present during winter. In these cases diapause seems usually to be associated with a high degree of "cold-hardiness." But cold-hardiness is not restricted to diapausing individuals. It may be developed by others that are quiescent without being in diapause. These forms seem capable of living almost indefinitely at moderately low temperatures; nor would it seem that they are readily killed by moderate exposures to extremes of low temperature, provided that they are not frozen. But this aspect of the subject has scarcely been studied, and there is little that can be said about it. It seems best to discuss the "cold resistance" of these dormant and cold-hardy forms in relation to the severity of the exposure required to freeze their tissues. This is a subject which has received much more attention.

6.313   THE LETHAL INFLUENCE OF FREEZING TEMPERATURES

It is characteristic of the water in the body fluids of animals that it tends to remain unfrozen, in a supercooled state, when the animal is chilled below 0° C. The strength of this tendency varies enormously in different animals, as we shall see below. The supercooling occurs because the water is present in a finely divided condition in association with colloids. For this reason and also because one organ or tissue may be quite different from another in this respect, the first outcome of chilling is almost certainly the formation of small isolated ice crystals here and there throughout the animal; doubtless there will be more in some tissues than in others. These, being insulated from one another and from the rest of the unfrozen water by the presence of colloids (which tend to become dehydrated in the vicinity of the ice crystals and thus enhance the insulation), grow slowly and tend not to "inoculate" the unfrozen supercooled water. Sooner or later, however, a threshold is passed or an "accident" occurs; the insulation breaks down somewhere; inoculation takes place and sets off a wave of freezing which sweeps through most parts of the animal's body (Salt, 1950). This is illustrated in Figure 6.25 (*solid line*). The curve illustrates the way the

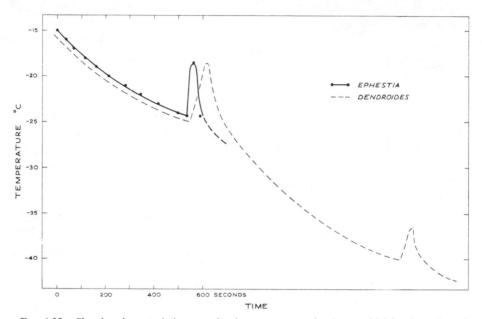

FIG. 6.25.—Showing characteristic curves for the temperature of an insect which has been abruptly transferred from room temperature to some very low temperature (e.g., −40 to −50° C.). Complete line for one that dies after the first rebound (based on data for *Ephestia* from Salt, 1936): Broken line is a hypothetical curve for an insect that survives the first rebound, as, for example, *Dendroides*.

temperature of an insect (*Ephestia*) changed when it was transferred abruptly from room temperature to a low temperature. It will be noticed that the temperature fell steeply and fairly smoothly to −24.3° C., when it abruptly increased to −18.6° and then proceeded to fall again. Supercooling ceased at −24.3° C.; the wave of crystallization which followed liberated enough heat to raise the temperature by 5.7°; then, as this heat was absorbed, the temperature began to fall again. The point at which supercooling came to an end is known as the "undercooling point"; the rise in temperature is the "rebound"; and the temperature to which the rebound rises has been called the "freezing point."

Usually the freezing that occurs during the rebound is sufficiently sudden and thorough to kill the animal; but exceptions to this rule also occur. It is not known what proportion of the total water in the tissues is frozen at the end of the rebound, for this is a matter that has received scarcely any attention. Almost certainly it will vary widely. Using the dilatometer method of Bouyoucos and working with fluids expressed from the bodies of a number of different insects, Sacharov (1930) measured the proportion of the total water in the body fluids that was frozen at different temperatures (see Table 6.15). These

TABLE 6.15*

PROPORTION OF BODY WEIGHT MADE UP BY WATER AND PROPORTION OF BODY WATER FROZEN AT
CERTAIN TEMPERATURES

SPECIES	WATER AS PER CENT LIVE WT.	ICE AS PER CENT TOTAL WATER				
		−3.9° C.	−5.8° C.	−7.8° C.	−11.1° C.	−17.4° C.
Euproctis larvae in diapause....	71.8	..	...	...	5.1	15.2
Euproctis larvae feeding.......	82.9	..	5.0	44.9	...	...
Euxoa larvae hibernating (autumn).................	71.4	..	27.0	...	...	...
Euxoa larvae hibernating (spring)....................	75.9	..	28.4	...	...	...
Euxoa larvae active (summer)..	84.7	..	67.4	...	...	...
Scoliopteryx adult in hibernation.....................	48.7	..	0.0	...	28.9	53.4
Plagionotus larva in hibernation	54.1	..	...	...	0.0	3.5
Locusta nymph active.........	86.2	0.0	20.9	...	...	...
Apis adult active............	74.0	..	37.0	...	73.9	...

* Data from Sacharov (1930).

figures may be quite different from what would obtain in the entire insect at the same temperature, but they indicate wide differences between species and stages. Evidence from experiments done in vivo is scarce, but a second undercooling point followed by a second rebound at a lower temperature has been observed on several occasions, from which it may be inferred that some water remained unfrozen after the first rebound. For example, Payne (1926b), working with a group of hardy beetle larvae (her oakborer group), observed the usual undercooling points and consequent rebound. The temperature at which the rebound occurred was highly variable but was above −20° C. for all the species. The insects apparently survived this wave of freezing, for all those that were thawed were found to be still alive. But if the insects were chilled still further, a second rebound was found somewhere below −40° C. This was always fatal (Fig. 6.25, *broken line*). Duval and Portier (1922) noticed that larvae of *Cossus* (Lepidoptera) had an undercooling point of about −12° C., but most larvae survived short exposures to −15° C. The larvae were chilled further, and in a few instances a second rebound was noticed at about −20° C. All larvae died after a short exposure to −21° C. Kozhanchikov (1938) observed that diapausing prepupae of *Croesus* (Lepidoptera) were "frozen" at −6° C.; yet he claimed to demonstrate a small amount of "thermostable" respiration down as low as −20° C. Since it is a priori unlikely that respiration would occur in the absence of liquid water, respiration at this temperature may be taken as evidence for the presence of unfrozen water at −20° C. in an insect that had appeared to be frozen at −6° C.

But the species that can be revived after they have experienced the first undercooling point and rebound are exceptional. With most species the freezing that accompanies the rebound is sufficient to kill them, and hence interest has centered around the first undercooling point. Most of the work on this subject has been based on the classical experiments of Bachmetjew (1907). He pushed a fine thermocouple junction into the tissues, usually the thorax of the insects with which he worked, and recorded its temperature as it was chilled. Bachmetjew identified the undercooling point as the lethal temperature, but he somewhat illogically considered that death was determined only when this temperature was reached for the second time, i.e., on the way down after the rebound. He also considered the temperature reached at the top of the rebound to be the "true freezing point" of the animal's tissues. This misconception has been perpetuated and has had a very wide currency, but a moment's reflection will show that it is largely determined by events which are quite unrelated to the nature of the tissues, i.e., to the temperature at the time the rebound starts, the amount of supercooled water that freezes during the rebound, and the weight, specific heat, and conductivity of the objects in the immediate vicinity of the animal when the experiment is being done. For a complex animal like an insect, the "freezing point" can hardly have a rational meaning; but if we like, we may attribute an arbitrary meaning to it and consider it the highest temperature at which a substantial amount of ice will form in the tissues when every effort has been made to reduce supercooling to a minimum. Whenever this has been done, the "freezing point" has been found to be very close to $0°$ C.—certainly not below $-2°$ C. (Salt, 1950).

The degree of supercooling that is obtained with a certain individual depends on a number of things. When its temperature is falling continuously, the spontaneous termination of the supercooled state becomes increasingly likely, the lower the temperature, down to about $-50°$ C.; beyond this limit the probability decreases with decreasing temperature (Luyet and Gehenio, 1940). But experience shows that the undercooling point is relatively independent of the speed of chilling, provided that this is relatively fast, i.e., fast enough for the undercooling point to be reached in minutes rather than hours (Salt, 1950). On the other hand, at a steady temperature the incidence of the spontaneous termination of the supercooled state becomes a function of time; the longer the animals are held at a constant (low) temperature, the more of them will experience the rebound. When a sample of 18 diapausing prepupae of the sawfly *Cephus* were held at $-20°$ C. for 120 days, they became frozen at the rate indicated in Table 6.16.

Since exposure to low temperature in nature has to be measured in hours or days rather than in minutes, the undercooling point which eliminates the influence of time is clearly a highly arbitrary concept, having only limited descriptive value (Salt, 1950). It follows that for an adequate description of the responses of animals to freezing temperatures, it is necessary to use the same methods that were discussed in the preceding sections in relation to non-freezing temperatures. That is to say, the lethal influence of a particular (freezing) temperature can be adequately described only in terms of the duration of the exposure required to kill the animal—and, for technical reasons, the best quantity to measure is the duration of exposure required to kill half of a random sample from the population. In this case it is possible to make the measurement directly on each individual, if it can reasonably be assumed that the moment the undercooling point occurs is always the moment at which death is inevitably determined.

TABLE 6.16*
TIME REQUIRED FOR "REBOUND" TO OCCUR IN DIAPAUSING PREPUPAE OF *Cephus* HELD FOR 120 DAYS AT −20° C.

Day	1	2	3	4	5	8	13	15	16	28	33	42	66	71
No. of larvae freezing	0	0	2	1	1	1	1	1	1	1	1	2	1	1

* On the 120th day, 4 were still unfrozen. Data from Salt (1950).

These are the fundamental limitations to the concept of the "undercooling point." They are reinforced by certain technical considerations: for example, the degree of undercooling may be greatly modified by prodding or jarring the animal or allowing it to wriggle, by the presence or absence of water in contact with the animal, and by wounding. Even the mild wound caused by inserting a fine thermocouple junction into the tissues may greatly raise the temperature of the undercooling point, so that results got by earlier workers using this method may not be compared with those of later workers with improved methods. Nevertheless, for special purposes these limitations do not matter. For example, the undercooling point, provided that it is measured in a standard way, may be a useful guide to the relative "cold-hardiness" of an individual.

6.314 COLD-HARDINESS

Kozhanchikov (1938) recognized that insects could be grouped into three classes according to their response to low temperatures. The ones in the first group cannot survive for any considerable period if the temperature falls below the lower limit of the range favorable for normal development. In other words, this sort cannot become dormant at low temperatures; they must either develop or die. This group is made up of (a) species that live in, or originated from, tropical or subtropical climates, e.g., *Calandra*, *Cimex*, and *Locusta* (see sec. 6.312); there are many of these species in which no stage in the life-cycle is capable of becoming dormant at temperatures below the favourable range; (b) species from temperate climates in which there is a certain stage in the life-cycle adapted to survive the winter but all other stages (i.e., those that are normally active during summer) resemble the subtropical and tropical forms in lacking the capacity to become dormant at low temperatures. Most members of this first group reach an undercooling point and rebound at a temperature only a few degrees below 0° C. There are exceptions, notably among those which are adapted to live in dry places or on dry food, e.g., *Calandra, Ephestia*. These may have undercooling points that are much lower. But this is not of much importance. All the members of this group, irrespective of their undercooling points, die when exposed to nonfreezing temperature; they are not "cold-hardy" in any meaning of this term.

The second group includes the forms that are able to become "quiescent" in the sense that Shelford (1927) used this term. That is, while remaining competent to develop at any time if they should be placed in the warmth, they may nevertheless survive, inactive but healthy, during prolonged exposure to cold. This group, popular belief to the contrary notwithstanding, is rare, and in searching the literature it is difficult to find many authentic representatives of it. Kozhanchikov (1938) placed the prepupal stage of *Agrotis segetum* (Lepidoptera) and the larval stage of *Lasiocampa quercus* in this group. The pupa of *Heliothis armigera* probably comes in here too.

The third group contains the diapause-stage of species from temperate climates. Many of the species that are adapted to live out-of-doors in temperate climates hibernate in a particular stage of the life-cycle; and frequently this stage and no other is capable of entering a state of diapause (sec. 4.2). The chief difference between members of the second and third classes is that the former become dormant directly in response to low temperature, but, with the latter, dormancy has usually arisen in response to some quite different stimulus (sec. 4.4). This may be primarily genetic and largely independent of environment; or, if environmental, the stimulus may have been operative at a much earlier stage in the life-cycle. The members of the two classes have in common a capacity to survive unharmed through quite prolonged exposures to nonfreezing low temperatures; usually they are killed only if their tissues are suddenly or extensively frozen, as, for example, during the rebound. For this reason, and in contrast with the first group, the members of the second and third groups might be said to be "cold-hardy"; but this term is usually reserved more specifically for the more cold-hardy members of these groups: those whose undercooling points are well below 0° C. The most cold-hardy of all may even survive the freezing that accompanies the first rebound and die only after a second wave of freezing, probably associated with a second undercooling point and rebound.

The degree of cold-hardiness, as measured by the undercooling point, differs widely between species. It may change in the one individual and usually does in rhythm with the season; or it may be modified artificially by desiccating the animal. There is a close correlation between the undercooling point and the amount of water in the tissues. This is brought out by the figures in Table 6.17, which has been compiled from data from Ditman *et al.* (1943), Payne (1926*b*,

1927*b*), Salt (1936), and Kozhanchikov (1938). There are exceptions to this rule. The water content of larvae of *Ephestia kühniella* changed not at all as it passed from the actively feeding stage to the prepupal stage, but the under-cooling point changed from −6° to −21° C. (Salt, 1936). Salt also found several other exceptions; but, as a rule, it is true that the lower the proportion of water in the body, the greater the cold-hardiness.

The adult beetle of *Leptinotarsa*, preparing for diapause, becomes dehydrated even in a moist, cool environment in the presence of plenty of lush food. This is

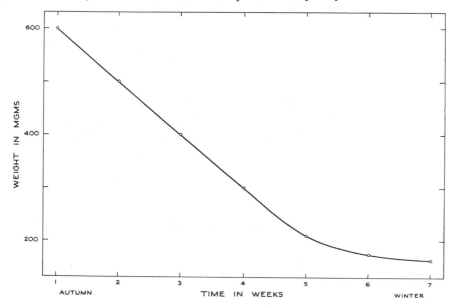

Fig. 6.26.—Showing the way in which the larva of *Diacrisia virginica* becomes dehydrated as it enters diapause at the close of the summer. (Data from Payne, 1927*b*.)

characteristic of diapause, whether it is a preparation for hibernation or for aestivation. In the former case dehydration is a valuable adaptation because thereby cold-hardiness is enhanced. The caterpillars of *Diacrisia virginica* at the close of summer weighed 600 mg. They entered diapause in this stage, and during the next 5 weeks lost weight (chiefly water) at the rate of about 80 mg. per week. As the live weight of the caterpillars approached 200 mg., the rate of loss slowed down greatly, and eventually the weight remained fairly steady at about 180 mg. (Fig. 6.26). The undercooling point became lower as the water-content was reduced, but the freezing associated with the first re-bound point was sufficient to kill the caterpillars up to the point where the weight-loss curve began to flatten out. But once the dehydration had pro-ceeded to the point where the live weight of the caterpillar was about 200 mg., the first rebound ceased to kill them; a much lower temperature (about −40° C.) was required.

The diapausing larva of the oakborer *Dendroides canadensis* becomes dehydrated in the same way, but not to the same extent, as *Diacrisia*. Figure 6.27 shows the water content during autumn and the seasonal trends in cold-hardiness in *Dendroides*. The "survival temperature" is the same as the undercooling point during September–October and May–June, but during winter it is very much lower; for *Dendroides* resembles *Diacrisia*, in that, once it has dried

TABLE 6.17*

RELATIONSHIP BETWEEN PROPORTION OF LIVE WEIGHT MADE UP BY WATER AND UNDERCOOLING POINT, FOR 11 SPECIES OF INSECTS

Species	Stage	Water as Per Cent Live Wt.	Undercooling Point (° C.)
Pyrausta	5th-instar larva, feeding	67–69	−11 to −15
	5th-instar prepupa, hibernating	56–57	−21 to −25
Anasa	Adult, feeding (summer)	63–68	−12 to −15
	Adult, hibernating (winter)	57–61	−17 to −23
Synchroa	Larva, summer normal	. . . . .	− 6 to −13†
	Larva, summer dehydrated (24 hr. over CaCl₂)	. . . . .	−12 to −26†
Popillia	Larva, hibernating normal	. . . . .	− 5 to − 7†
	Larva, hibernating, dehydrated to 50% of original water content	. . . . .	*ca* − 28†
Musca	Puparia, normal	. . . . .	−12 ± 3.2
	Puparia, dehydrated to lose 10% of original weight	. . . . .	−22 ± 1.8
Agrotis	Larva, feeding	88	− 2 to − 3
	Prepupa, hibernating	73	−10 to −11
Loxostege	Larva, 4th-instar, feeding	85	− 2 to − 3
	Prepupa, 5th-instar, hibernating	60	Lower than −20
Croesus	Prepupa in diapause	61	Lower than −20
Acronicta	Pupa in diapause	63	Lower than −20
Lymantria	Eggs in diapause	62	Lower than −20
Cydia	Prepupa in diapause	59	Lower than −20

* Data from Ditman *et al.* (1943), Payne (1926*b*, 1927*b*), Salt (1936), Kozhanchikov (1938).
† Results probably got by piercing tissues with thermocouple junction.

out beyond a certain point, it is able to survive the freezing associated with the first rebound and dies only after a second undercooling point has occurred; this is usually about −40° (Payne, 1926*a*, 1927*b*). In northeastern United States the corn earworm, *Heliothis armigera*, hibernates as a pupa. Barber and Dicke (1939) exposed a number of pupae out-of-doors; some were in dry soil, and some were in moist soil. The death-rates in the two groups are shown in Table 6.18. Since the moist soil was a better conductor than the dry, these

TABLE 6.18*

DEATH-RATES IN TWO GROUPS OF PUPAE OF *Heliothis armigera* EXPOSED OUT-OF-DOORS DURING WINTER

EXPERIMENT	DEATH-RATE AS PER CENT	
	Moist Soil	Dry Soil
First	27.4	0.0
Second	52.0	1.3

* The experiment was repeated during a second winter. After Barber and Dicke (1939).

pupae were probably colder than those in the dry soil. But the difference between the two groups may have been due to the failure of the pupae that were kept moist to become cold-hardy. Hodson (1937) found that the bug *Leptocoris trivittatus* did not become more cold-hardy when it was desiccated; and Sweetman (1929) found the same with the beetle *Epilachna corrupta*. These may be regarded as exceptions to the general rule.

One outcome of the seasonal fluctuation in cold-hardiness, which is illustrated in Figure 6.27, is that when some constant test (for example, ex-

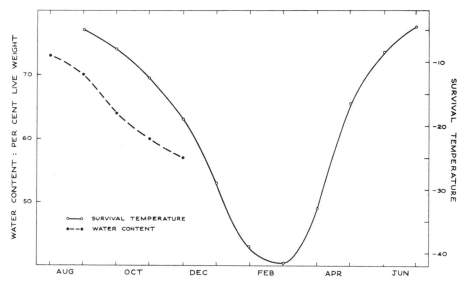

Fig. 6.27.—Showing the falling water content (*broken line*) of larval *Dendroides canadensis* as it enters diapause in the autumn, and the seasonal trend in cold-hardiness (*complete line*) of larval *Dendroides*. (Data from Payne, 1927b.)

posure to $-20°$ C. for 3 hours) is applied to samples of the population at different seasons of the year, different proportions of the animals in the samples will become frozen at different seasons. For example, if this test were applied to, say, 100 *Dendroides* in September, most of them might experience an undercooling point and rebound and become frozen; but if the same test were repeated in December, perhaps the proportion freezing would be less than half of the total number in the sample. An individual which happened not to freeze in September would contain about the same low amount of ice in its tissues as one that did not freeze in December; and one that froze in December would contain about the same large amount of ice as one that froze in September. But because many more individuals in the sample of 100 froze in September than in December, there would be, in the aggregate, more ice present in the September sample. This is all very elementary, and it must surely be clear that no useful purpose is served by coining the misleading term "bound water" to describe

that portion of the supercooled water which happens to be located in those individuals which happen not to have experienced an undercooling point and rebound. Yet this is just the experiment that Robinson (1928) did with *Cal-
losamia* and *Phyllophaga*, which has been so widely and uncritically quoted as
evidence that cold-hardiness is due to the presence of "bound water." Now that
Kistler (in Salt, 1936) has clearly shown that "bound water" in Robinson's
experiments was no more than supercooled water—and, indeed, a moment's
thought along the lines indicated in this present paragraph will make this quite
clear—there seems to be no reason for allowing this misleading conception to
persist.

6.315   SURVIVAL-RATE IN NATURE IN RELATION TO FREEZING TEMPERATURES

The reduction in cold-hardiness in the spring (Fig. 6.27) is associated with
the completion of diapause-development and the consequent increase in water
content which is characteristic of the post-diapause-stage. This is a reason why
a population which has withstood great extremes of cold during winter may
suffer high mortality during a late spring frost, when the temperature is much
less extreme than it was during winter. As an example we refer to the observations of Payne (1926*b*), who studied a population of oakborers which during
winter may withstand temperatures as low as $-40°$ C. In the fall the undercooling point usually goes down ahead of the temperature, acting as a factor of
safety. In the spring when the insects have begun to lose their hardiness, a sudden cold snap or a blizzard is fatal to many of them. During 1924 there was a
sudden cold wave during the first part of the month, with a blizzard on April 1.
After this, about 43 per cent of the larvae were found to be dead. If hazards of
this sort were severe and frequent, the individuals that were tardy in complet-
ing diapause would be favored. But it so often happens that the favorable
period for growth and reproduction is short during spring and early summer, so
that selection pressure may also be operating in the other direction.

The northern limit of distribution of a species may often be determined by
low temperatures. A striking example, which also illustrates the point made
in the first chapter that distribution and abundance are but different aspects
of the same phenomenon, is provided by the moth *Heliothis armigera* in eastern
North America. The species is a serious pest of corn and cotton in the southern
states, where it is well established and maintains relatively high numbers; in
the absence of any important natural checks, the numbers are kept down only
by routine destruction of the caterpillars with insecticides. Farther north in the
region of New York, most overwintering pupae are killed by cold. The survivors have their numbers augmented during summer by immigrants from
farther south; but, on the whole, the species is less abundant than it is in the
southern states (Barber and Dicke, 1939). Farther north still, in Canada, it is
said that every pupa is killed every year and that the summer population is

derived entirely by immigration from farther south. The numbers fluctuate sporadically and are never very high (Pond, 1948). Sacharov (1930) considered that the distribution of the cutworm *Euxoa segetum* as a pest in the region of the lower Volga may be largely explained in terms of the influence of freezing temperatures on overwintering larvae. The species is permanently established throughout the region, but it multiplies to become a pest quite infrequently, and then only in certain areas which are characterized by a thicker and more reliable covering of snow. Elsewhere, where the cover of snow is thin or lacking, temperatures in the soil at the depth where the larvae overwinter are generally so low that most larvae are destroyed most years and the population never gets a chance to increase to large numbers. Sacharov observed the rise and decline of one outbreak. Unusually heavy falls of snow during several winters prior to 1924 permitted the numbers to increase in several areas in the region of the lower Volga. The outbreak came to an end during the winter of 1925–26, when the cover of snow was unusually thin everywhere and temperatures of the order of $-12°$ C. were registered in soil in places where the overwintering larvae were numerous. These temperatures were fatal. This is just another instance of the way in which distribution and abundance are under the control of the same sorts of influences.

With a species that is well adapted to withstand cold, low temperature may not be an important hazard over most of its distribution. This would seem to be the case with *Cephus* in the Canadian prairie; for Salt (1950) reported that, of the many thousands that he had collected from the field over a period of years, he had not found one that had been killed by the cold of winter. Similarly, with the cornborer *Pyrausta*, Stirrett (1931, 1938) showed that the death-rate during winter in Ontario was negligible, rarely exceeding 10 per cent. This was because the temperature during winter in the corn-growing areas of Ontario was rarely, if ever, low enough to kill the cold-hardy diapausing fifth-instar larva of *Pyrausta*. As we have seen, a cold-hardy stage is characteristic of many species of temperate climates. Nevertheless, extreme fluctuations in weather may occasionally cause many deaths, even in species that are usually secure.

The beetle *Dendroctonus brevicornis* lives in the bark of several sorts of pines in northwestern North America, in areas where freezing temperatures occur in the winter. The insect is adapted to withstand cold, and the death-rate during most winters is small. But during the winter of 1924–25 unusually low air temperatures of the order of $-27°$ C. were recorded, and on this occasion from 25 to 80 per cent of the *Dendroctonus* in the bark were killed by cold. An outbreak of the related *D. frontalis* in West Virginia came to an end after the severe freeze of 1892–93 (Miller, 1931).

A high death-rate during winter may be the rule rather than the exception. The bug *Perillus bioculatus* preys upon the Colorado potato beetle, *Leptinotarsa decemlineata*, and in certain circumstances may be important in reducing its

numbers. Knight (1922) during 7 years' observations of *Perillus* in New York
and Minnesota observed a high death-rate during winter each year.

> Under natural conditions of hibernation the mortality of the bugs must be very high dur-
> ing most years. It has been the writer's experience during six or seven years' observation
> in the potato fields in New York and Minnesota that very few bugs appear in the fields
> the following spring where in the preceding fall they were abundant. Under natural condi-
> tions it appears that probably not more than five per cent. of the bugs that go into hiberna-
> tion come out safely in the spring. In New York and Minnesota this may well be due to
> the fact that many of the bugs seek hibernation in situations where their fatal minimum
> temperature is reached during the cold winters. In the spring of 1921 the largest number
> of bugs appeared in the field that the writer has ever observed. The fall of 1920 was very
> favorable for the bugs that went into hibernation, and following this the winter was unu-
> sually mild. The mild winter probably allowed many bugs to hibernate safely in situations
> where during ordinary cold winters their fatal temperature would have been reached. Such
> conditions would seem to account for the greater abundance of the bugs in the spring of 1921

Knight did not add, though he might have, that during an unusually cold
winter the converse would hold and only those very few bugs which had happened to find especially safe situations would survive. This is an example of the
interaction between the animal's "place to live" and other components of environment (in this case low temperature) which we discuss in section 12.2.
This principle is of great importance in understanding the distribution and
abundance of animals in nature. Salt (1950) summarized this principle neatly.
Insects that cannot survive the formation of ice in their tissues "depend for
their survival on the insulating protection of their hibernacula (soil, plant
debris, snow, ice, etc.) and on their ability to undercool. The combination of
these two protective factors may be sufficient that mortality as a result of
freezing is a rarity in some species. More often the protection is incomplete,
with the result that those individuals in the more exposed hibernacula and with
the least ability to undercool perish. In severe winters this fraction may be
temporarily large and may seriously deplete the population. If a species has
been expanding its range into colder regions, one severe cold period can eradicate it in such areas, restricting it once again to its normal range."

6.32   *The Lethal Influence of High Temperatures*

There has been much less work done on the lethal influence of high temperatures than of low temperatures. With terrestrial animals it may be difficult to
devise an experiment which adequately measures the influence of lethal high
temperatures independently of the influence of moisture. Moreover, it is often
more interesting to study the influence of moisture because, in nature, harmful
high temperatures are most likely to occur in deserts or in warm, temperate
zones where the summer is arid. In such places the danger from evaporation is
likely to be more pressing than the danger from heat. With aquatic animals,
especially with certain species of fishes which live in shallow lakes or rivers in
temperate zones, high temperatures may sometimes cause a lot of deaths. Two

aspects of the lethal influence of temperature on fishes have received special attention, namely, the limits of the tolerable zone and the influence of acclimatization on the "heat resistance" of the animal.

6.321   THE LIMITS OF THE TOLERABLE RANGE

If an animal is exposed to an extreme high or an extreme low temperature, it will quickly die; and no practical difficulties are raised when we say that death was caused by the exposure to unfavorable temperature. As the temperature becomes somewhat less extreme, death may approach more slowly, but it may still be clear that exposure to unfavorable temperature was the chief cause of death. Moving farther and farther away from the extremes, we come to a zone of moderate temperature where it is not reasonable to attribute any lethal influence to temperature. Fry (1947) called the two limits to this median tolerable zone the "incipient lethal low temperature" and the "incipient lethal high temperature"; and he went on to say that above and below the median tolerable zone are the "upper lethal zone" and the "lower lethal zone," respectively (Fig. 6.28). Can we discover the limits of the median tolerable zone by experiment?

In order to discover these limits, we need to know the least extreme temperature which would be fatal. Since the time required for a fatal exposure increases as the temperature becomes less extreme, this means that we need to know the least extreme temperature at which the time required for a fatal exposure becomes indefinitely long. When we come to do the necessary experiment, we find that, as the limit is approached, the duration of exposure becomes so long that eventually we may be unable to decide whether the animal died of old age or from some harmful influence of temperature. Thus there is no empirical way to measure these limits precisely. In practice, it is necessary to substitute arbitrarily some definite period for the "indefinitely long" exposure required by the theory. For example, in their experiments with young speckled trout (*Salvelinus*) Fry *et al.* (1946) used periods that varied from 1,200 to 5,000 minutes, depending on the circumstances. But in each case they chose a period that was sufficiently long to justify the assumption that any fish which was still alive at the end of this period would have been able to survive an indefinitely long exposure to the particular temperature of the experiment. In this way an approximate result can be obtained.

6.322   ACCLIMATIZATION

The limits of the tolerable zone are not "fixed" but depend to some extent upon the condition of the animal. For example, it has been shown, especially with fish, that the temperature at which an animal has been living may have quite a profound influence on the limits of the tolerable zone. This effect is usually called "acclimatization." Figure 6.28, which has been constructed from the data of Fry *et al* (1942) for the goldfish, shows that the tolerable range moves up the temperature scale and becomes narrower, the higher the temperature at which the fish had been living before the experiment began. There is an upper and a lower limit to the temperatures at which acclimatization goes on, and there is a corresponding limit imposed on the variation in the

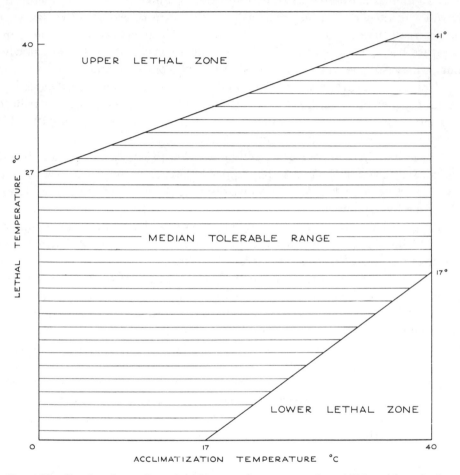

FIG. 6.28.—Showing the median tolerable range of temperature for goldfish and how the lower and upper lethal limits depend upon the temperature to which the fish had previously been acclimatized. (Modified after Fry, Brett, and Clausen, 1942.)

incipient lethal temperatures which can be brought about by acclimatization. Fry (1947) considered these limits to be characteristic of the species. Thus with goldfish the acclimatization effect is not accentuated by exposures above 41° C., and the limit above which the incipient lethal low temperature cannot be raised by acclimatization at high temperature is 17° C. (Fig. 6.28). Taking into account the full potential influence of acclimatization, the lowest incipient lethal low temperature for the goldfish may be below 0° C., and the highest incipient lethal high temperature may be near 40° C.

Acclimatization influences not only the upper and lower incipient lethal temperatures but also the duration of exposure to a lethal temperature which an animal may withstand. Figure 6.29, *A*, illustrates this. It has been constructed from Fry *et al.*'s (1946) data for yearling speckled trout. The diagram indicates clearly that the duration of exposure to lethal high temperature re-

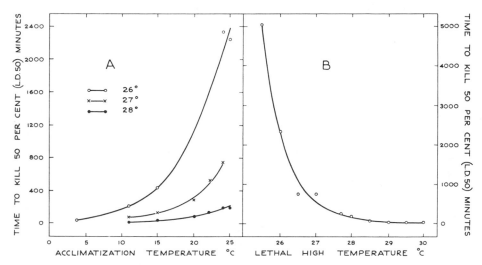

FIG. 6.29.—*A*, showing the influence of acclimatization on the survival time of trout, exposed to three lethal temperatures, 26°, 27°, and 28° C. The influence of acclimatization was most pronounced when the lethal temperature was 26° C. But at all three temperatures fish which had been acclimatized at a high temperature lived longer than those which had been living at a lower temperature. *B*, showing the survival time for trout exposed to lethal high temperatures ranging from 25.5° to 30° C. The fish had all been living at 24° C. before the experiment began. (Data from Fry, Hart, and Walker, 1946.)

quired to kill half the fish was less for those that had been living at a low temperature than for those that came from the higher temperature.

In nature, acclimatization occurs from one season to the next, as was well shown by Brett's (1944) data for bullhead (*Ameiurus*) living in Lake Opeongo, Ontario. He collected fish from the lake at intervals from July, 1940, to September, 1941, and measured the high temperature required to kill half the sample during an exposure of 12 hours. Figure 6.30 shows that the seasonal trends are most striking. The capacity to become acclimatized to changing levels of temperature is a useful adaptation for fish which live in lakes and rivers in a temperate climate. In Ontario the extremes of temperature are such that fish inhabiting these situations occasionally suffer severe mortality from the direct influence of high temperatures (Huntsman, 1946). If they lacked this adaptation, the death-rate would doubtless be higher, and the occasions on which many died would happen more frequently, with the consequence that the average numbers in the area might be lower. One might expect similar phenotypical differences to occur between populations from different climatic zones, but this may be complicated by the existence of genetic differences in such populations. Hart (1952) compared the lethal temperatures of fresh-water fish from different localities in North America. He studied 14 species. Geographic differences in the upper lethal temperature were found only in three species—*Notropis cornutus*, *Gambusia affinis*, and *Micropterus salmoides*, each

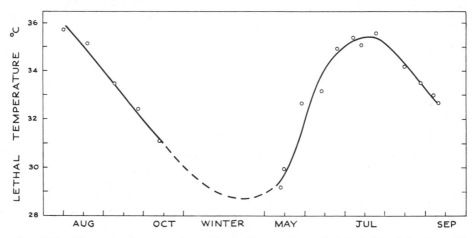

Fig. 6.30.—Illustrating the seasonal trend in the "heat resistance" of the bullhead *Ameiurus*. The ordinate gives the lowest temperature at which just half the sample of fish were dead after an exposure of 12 hours. The fish have clearly become acclimatized to the prevailing temperature of each season. (After Brett, 1944.)

from southern Canada and Tennessee. In each case he was able to find morphological differences between the populations from the different regions, and he recognized them as subspecies. Hart suggested that the similarity in the upper lethal temperatures for the other 11 species may have been due to their great capacities for acclimatization; or perhaps it was due to the relatively low temperatures prevailing in the places where these fishes lived. Since none of them ran any risk of being killed by high temperature, there was little likelihood of any of them being selected for this character.

6.323   DELAY IN THE MANIFESTATION OF THE LETHAL INFLUENCE OF HIGH TEMPERATURE

We mentioned in section 6.31 that it is often not possible to tell merely by looking at an animal the precise moment when it has had a fatal "dose" of high or low temperature. Sometimes the harm that has been done cannot be detected until much later in the life-cycle. This is well illustrated by Darby and Kapp's (1933) experiments with maggots of the fruit fly *Anastrepha*. In nature the maggots of *Anastrepha* live surrounded by the juicy contents of the fruit, so it is possible to measure the influence of high temperature on survival-rate, independent of the influence of moisture. Darby and Kapp exposed batches of larvae in their third instar to a series of temperatures between 36° and 46° C. for 4 hours. The deaths from heat were negligble up to about 40°; an exposure of 4 hours to 42° killed about half of the larvae; and at 46° all were dead at the end of 4 hours. In another experiment, batches of larvae were exposed to 40.5° C. for various periods. About 11 hours at 40.5° were required to kill about half the larvae; and all were dead after 15 hours.

Darby and Kapp not only recorded the number of deaths at the end of the 4 hours' exposure to the various high temperatures but also gave the survivors the opportunity to continue their development and pupate at a favorable temperature. Most of them did complete their larval development, but relatively few of them were able to pupate. Many died at this stage. The percentage of surviving larvae which also survived to become pupae is shown in Table 6.19. It is clear that exposure to high temperatures did a certain amount of harm which could not be measured by counting those which died immediately. A large proportion of those which apparently recovered were nevertheless unable to complete their development, dying at the next "critical" stage in their life-cycle. This "delayed-action" effect of high temperature is well known to students of this subject. Often the survivors appear quite healthy at first and may feed and grow apparently quite normally until some critical stage, e.g., ecdysis or metamorphosis, is reached. Then the harmful influence of the previous exposure to high temperature becomes manifest in the unusually high death-rate at this stage (Larsen, 1943).

TABLE 6.19*
DELAYED HARMFUL INFLUENCE OF HIGH TEMPERATURE ON LARVAE OF *Anastrepha*

Temperature (° C.)	Per Cent Pupae
40.8	53
41.4	49
41.8	41
41.9	13

* The figures show the proportion of larvae which, having survived exposure to high temperature for 4 hours, were subsequently able to pupate. After Darby and Kapp (1933).

6.324   THE "HEAT RESISTANCE" OF ANIMALS THAT LIVE IN WARM PLACES

In section 6.233 (and see Fig. 6.01) we showed how animals that normally inhabit warm places are adapted to develop at higher temperatures than those from cold places. Similarly, as might be expected, the capacity to survive exposure to extreme temperatures is related to the temperatures which prevail in the places where the animal lives. For example, the nymphs of the firebrat *Thermobia domestica*, which normally inhabits such hot places as the hobs of bakers' ovens, may live indefinitely at temperatures as high as, or higher than,

42° C. But the nymphs of the related silverfish, *Lepisma saccharina*, which, being a common household pest in temperate climates, inhabits much cooler places, are unable to withstand temperatures above about 36° C. (Sweetman, 1938, 1939). Buxton (1924) found certain species of Orthoptera living in the desert in Palestine on bare earth with a temperature of 60° C.; but it is unlikely that even these desert forms could survive for very long at such a high temperature. Chapman *et al.* (1926) measured the temperature in a number of situations around a sand dune in Minnesota. On a bright summer's day the temperature on the surface of the sand was 51° C., which is comparable with that found in the desert. But at the same time the temperature of the air about 12 inches above the surface of the sand was 27° C., and that of the sand 12 inches below the surface was 38° C. On a cloudy day with rain falling, the temperature of the air 12 inches above the surface was 18° C., and that of the sand 12 inches below the surface was 22° C. The fauna of the dune included the sand wasp *Bembex* and its parasite *Dasymutilla*. The former constructs deep burrows in the sand and stocks them with flies, which serve as food for the larvae. The latter seek out these burrows and lay their eggs on the larval *Bembex*. The adult *Bembex* was inactive at temperatures below about 25° C., so that its activity was largely restricted to bright days. Yet it succumbed rapidly to high temperatures above about 42° C. Nevertheless, it spent a substantial part of its time on bright, sunny days burrowing in the sand. It contrived to get through the surface layers, where the temperature of the sand was so high that exposures of more than a few minutes would be fatal to the wasp, by short bursts of digging interspersed with frequent visits to the layers of air above the dune where the temperature was tolerable. But its parasite, *Dasymutilla*, being wingless, might not evade the temperature of the surface sand except by entering the burrows of *Bembex* or by climbing the sparse vegetation on the dune. Neither device affords adequate protection, as the insect must spend a lot of time crawling on the surface of the sand. It is able to survive on the dune because it is adapted to withstand high temperature. Chapman *et al.* (1926) found that when a batch of *Dasymutilla* females were gradually warmed during 160 minutes from 0° to 56° C. the first sign of heat stroke appeared at 52°, and all the insects were knocked over by the heat when the temperature reached 55°; in a comparable experiment with adult *Bembex*, the corresponding temperatures were 42° and 44° C. These authors also observed: "In the normal course of the day all the insects leave the surface of the sand when its temperature nears 50° C. Some climb grasses and some enter their burrows, while others fly about some distance above the sand making hurried landings to enter their burrows. The female mutillids were consistently the last to retreat when the temperature rose and the first to return to the open sand when the temperature fell."

6.4 SUMMARY

We have seen in the preceding sections how temperature may influence behavior, speed of development, and longevity and that acclimatization may be important with respect to all these aspects. Leaving behavior aside for the moment, it is possible to epitomize the remainder in a single hypothetical diagram. This has been done in Figure 6.31. The zone of lethal low temperature has been divided into two subzones; A_1 is the zone of freezing temperatures, which is chiefly important in relation to forms that can become dormant, and, for the most part, this means diapausing individuals: A_{11} is the zone of nonfreezing lethal low temperature, which is chiefly important with respect to nondormant forms. In zones A_1 and A_{11} and in zone C the influence of acclimatization is indicated by drawing the curves as a band rather than a line. In both zones A and C, interest centers on the lethal influence of temperature, and this is expressed as the duration of exposure required to kill half the sample of animals (L.D. 50). This approaches infinity at the incipient lethal temperatures, i.e., at the limits to the zone of favorable temperature, and approaches zero at extremes of low temperature A or high temperature C. The central zone is the zone of favorable temperature, and here interest centers on the speed of development which is represented in the diagram by portion of a logistic curve.

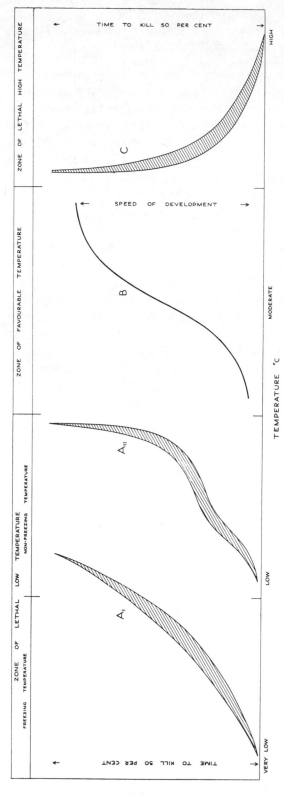

Fig. 6.31.—Hypothetical curves summarizing the influence of temperature on the animal throughout the whole range of temperature of interest in ecology. In zones *A* and *C*, interest centers on the lethal influence of temperature and is expressed as the duration of exposure that is lethal. In *A*, separate curves are drawn for hardy and nonhardy forms. The curves are drawn as bands rather than lines, to indicate the influence of acclimatization. In zone *B*, interest centers on speed of development, and this has been expressed as part of a logistic curve. For further explanation see text.

CHAPTER 12

A Place in Which To Live

We have three butterflies which are limited by geological considerations, being inhabitants only of chalk downs or limestone hills in south and central England, and they may reach the shore where such formations break in cliffs to the sea. These are the Silver-spotted Skipper, Hesperia comma, *the Chalk-hill Blue,* Lysandra coridon, *and the Adonis Blue,* L. bellargus. . . . *The two latter insects are further restricted by the distribution of their food plant, the Horse-shoe Vetch,* Hippocrepis comosa, *and possibly by the occurrence of a sufficiency of ants to guard them. Yet any of the three species may be absent from a hillside which seems to possess all the qualification which they need, even though they may occur elsewhere in the immediate neighbourhood. This more subtle type of preference is one which entomologists constantly encounter, and a detailed analysis of it is much needed. A collector who is a careful observer is often able to examine a terrain and to decide, intuitively as it were, whether a given butterfly will be found there, and that rare being, the really accomplished naturalist will nearly always be right. Of course he reaches his conclusions by a synthesis, subconscious as well as conscious, of the varied characteristics of the spot weighed up with great experience; but this is a work of art rather than of science, and we would gladly know the components which make such predictions possible.*

FORD, *Butterflies* (1945a, p. 122)

When two sticklebacks meet in battle, it is possible to predict with a high degree of certainty how the fight will end: the one which is farther from his nest will lose the match. In the immediate neighbourhood of his nest, even the smallest male will defeat the largest one. . . . The vanquished fish invariably flees homeward, and the victor, carried away by his successes, chases the other furiously, far into its domain. The further the victor goes from home, the more his courage ebbs, while that of the vanquished rises in proportion. Arrived in the precincts of his nest, the fugitive gains new strength, turns right about and dashes with gathering fury at his pursuer. A new battle begins which ends, with absolute certainty, in the defeat of the former victor, and off goes the chase again in the opposite direction. The pursuit is repeated a few times in alternating directions, swinging to and fro like a pendulum which at last reaches a state of equilibrium at a certain point. The line at which the fighting potentials of the individuals are thus equally balanced marks the border of their territories. The same principle is of great importance in the biology of many animals, particularly that of birds.

LORENZ, *King Solomon's Ring* (1952, pp. 26–27)

12.0   INTRODUCTION

A GOOD case might be made out for the proposition that among vertebrates the behavior patterns associated with territory are more fundamental even than those associated with sex. With invertebrates, "territoriality" as

103

such hardly occurs, but the behavior associated with the animal's choice of a place to live seems to be deeply rooted and complex. So it is probably true of animals in general that the seeking of a special place to live in is one of their most fundamental characteristics. Since this character, like all others, is subject to natural variation within a population and since any area which may support a natural population is inevitably variable with respect to the sorts of places where the animals may live, it follows that some individuals are always better placed than others not only with respect to the usual hazards associated with weather, other animals, food, and so on but also with respect to many less tangible requirements which the naturalist can only guess at.

The quotation from Ford which appears at the head of this chapter was written from the point of view of the collector; but the ecologist who wishes to have a quantitative appreciation of the distribution and abundance of the species that he is studying must also strive to emulate "that rare being, the really accomplished naturalist"; he must learn the habits and requirements of the species that he studies, so that he can say for each one what are the sorts of places it can live in and why. In this chapter we shall discuss this problem from three aspects: for want of a better way to discriminate between them, we call these "relative," "absolute," and "quantitative" aspects. By the "relative aspect" we mean that certain sorts of places may be valuable chiefly because they provide shelter against the extremes of weather, or protection against a predator, or so on; in other words, the way they influence the animal's chance to survive and multiply depends on the influence of other components in the environment. This is discussed in section 12.2. On the other hand, the suitability of a certain sort of place may depend less on its interaction with other components of environment than on its own innate qualities and the behavior of the animal. This is discussed in section 12.1. But irrespective of whether the places where animals may live are chiefly important for their relative or their absolute qualities, it is always important to know how numerous such places may be and how they may be distributed in a particular area. This is what we mean by the "quantitative aspect," and we discuss it in section 12.3.

12.1   SOME EXAMPLES OF THE PLACES WHERE ANIMALS MAY LIVE AND THE SUBTLE BEHAVIOR OF ANIMALS IN CHOOSING A PLACE TO LIVE

Elton (1927, p. 39) wrote: "Most animals have some more or less efficient means of finding and remaining in the habitat which is most favourable to them . . . most animals are, in practice, limited in their distribution by their habits and reactions, the latter being so adjusted that they choose places to live in, which are suitable to their particular physiological requirements or to their breeding habits." As an example he mentioned how the African lion chooses its lair with great attention to a number of rather subtle requirements.

At the other extreme there are numerous insects and other small animals which merely produce a great superabundance of individuals with a powerful urge for dispersal. They launch themselves into the air and are wafted wherever the wind may blow (sec. 5.13). The extent of their freedom to choose a place in which to live would seem to lie in their capacity to recognize a good place when they happen to have alighted on it and to stay there if the place satisfies their requirements. The aphids, scale insects, and mites do this. The reputations which some of these species have as pests of horticultural crops bear witness to the efficacy of this means of finding a place to live. But most animals display more complex behavior than this in seeking a place in which to live.

The requirements of the small case-bearing moth *Luffia ferchaultella* (Psychidae) have been worked out fully by McDonogh (1939). It may be found in southern England, living on the trunks of trees and on stones where lichen is growing. The limits of its distribution are near to the 62° F. isotherm for July. It is found in some places where the mean temperature for July is below 62° F., but these are places with an unusually large amount of sunshine. It is absent from some places where the mean temperature for July is 62° F. or higher, but these are mostly places where the temperature during winter is unusually low. Within the area bounded by this isotherm, *Luffia* occurs from sea-level up to an altitude of about 400 feet, but not in higher places. Outside these limits there are trees and stones which seem in every other way suitable for *Luffia*, but none occurs there, presumably for some reason associated with temperature. McDonogh's explanation for the northern limit was that places where the mean temperature for July is below 62° F. are likely, during winter, to experience temperature low enough to kill *Luffia*, even those individuals which had found the best shelter.

The caterpillars thrive when they have as food the lichen *Lecanora*, which is composed of a green alga, *Pleurococcus*, and the hyphae of a fungus which was not identified. Any small area of bark chosen at random might have growing on it (*a*) neither *Pleurococcus* nor *Lecanora*, (*b*) either one or the other predominantly, or (*c*) both patchily. McDonogh counted the numbers of larvae in November and again in April on patches of bark supporting weak and strong growths of the alga and the lichen. In Table 12.01 the results are given as the average number of larvae on 6 square inches of bark. Although the caterpillars were present in low numbers in places where there was only *Pleurococcus*, they were numerous only in places where *Lecanora* predominated and was growing densely.

There were, however, certain sorts of places where *Lecanora* was growing densely which were characteristically devoid of *Luffia*. The height above ground-level and the position of the tree with respect to the edge of the wood are important. On the latter point McDonogh wrote: "The larvae are not found on trees in a wood if the tree is more than 20 yards from the edge, unless the

TABLE 12.01*
AVERAGE NUMBER OF LARVAE OF *Luffia ferchaultella* FOUND ON SAMPLE AREAS OF
6 SQUARE INCHES OF BARK ON HORSE CHESTNUT WHEN THESE SUPPORTED FOOD IN
VARYING AMOUNTS AND OF VARYING QUALITY

NATURE AND QUALITY OF FOOD	AVERAGE NO. OF LARVAE IN 6 SQUARE INCHES		
	November	April	Mean
60% *Lecanora*, thick growth................	2.2	2.8	2.5
80% *Lecanora*, patchy growth...............	2.0	2.4	2.2
90% *Lecanora*, thick growth...............	5.6	5.0	5.3
No *Lecanora*.............................	0.3	0.7	0.5
80% *Pleurococcus*, thick growth...........	0.1	0.0	0.1
Both types present but scattered............	0.3	0.6	0.5
Both types present and thick...............	0.5	0.6	0.6

* Data from McDonogh (1939).

undergrowth is very thin. Trees surrounded by thick bushes do not have the
moth on them. . . . Besides this screening effect due to trees and undergrowth there is another caused by sudden rises in the ground-level. A group of
trees situated on the top of a hill will more often than not be uninfested by the
moth, though the trees are apparently suitable. . . . Gentle undulations do
not affect the distribution. The optimum type of country for the moth is open
park land such as at Richmond Park or Windsor Great Park, where there are
plenty of trees situated in unscreened positions."

McDonogh also counted the number of larvae on the trunks of horse chestnut trees at various heights above the ground. The figures in Table 12.02 refer

TABLE 12.02*
MEAN NUMBER OF LARVAE OF *Luffia ferchaultella* ON TRUNK OF HORSE CHEST-
NUT TREES AT VARIOUS HEIGHTS ABOVE GROUND

HEIGHT ABOVE GROUND	MEAN NO. OF LARVAE PER 6-INCH SQUARE				
	North	East	South	West	Mean
6 inches..........	3.75	3.75	19.75	3.00	7.56
2 feet............	0.75	2.50	7.50	0.75	3.00
4 feet............	1.50	2.00	5.00	0.50	2.25
5 feet............	0.50	1.75	3.00	0.75	1.50
Mean........	1.63	2.50	8.81	1.30	3.58

* Data from McDonogh (1939).

to the mean number of larvae on 36 square inches of bark that was well
covered by *Lecanora*. The larvae were most abundant near the ground and on
the south side, and scarcely any were found more than 8 feet above the ground.

The absence of *Luffia* from the upper reaches of the trunk and from trees
which are deeply shaded may be explained by their reactions to light. Mc-
Donogh's experiment with light is best described in his own words: "A set of
green screens was made to cover part of a tree trunk, so that the light intensity
over that part of the surface was reduced. They were in three degrees of density
and are referred to here as light, mid- and dark green screens. At the beginning
of each experiment a known number of active larvae was placed under each of

the screens. After 24 hours the position of the larvae under the screens was noted. Two areas 4 and 6 inches square were marked under the centre of each screen. The number of larvae found in the squares was taken as a measure of the effect of the screens on the movement of the larvae compared with the movement of a similar set of larvae in a control area without the screens." The results of this experiment are given in Table 12.03.

TABLE 12.03*
NUMBER OF LARVAE OF *Luffia ferchaultella* (AS PER CENT OF TOTAL) WHICH REMAINED IN EACH LEVEL OF ILLUMINANCE 24 HOURS AFTER BEGINNING OF EXPERIMENT

ILLUMINANCE (AS RATIO OF FULL LIGHT IN OPEN)	TYPE OF SCREEN	NO. OF LARVAE (AS PER CENT OF TOTAL)	
		4-In. Square	6-In. Square
0.023	Dark green	7.4	37.2
.047	Mid-green	31.8	56.2
.070	Light green	37.8	60.4
0.117	None (control)	26.8	56.2

* Data from McDonogh (1939).

The elm hardly ever provides a favorable place for *Luffia* to live. The willow and the oak nearly always harbor at least a few, and often quite dense populations live on them. Sometimes the beech and the horse chestnut also provide good places for *Luffia* to live. Young trees of any species which have not attained a girth of about 8 inches rarely support many *Luffia*. The differences between species and the inadequacy of young trees may be associated with the requirements outlined above, but there is also the matter of crevices which are required by the pupae. "While the maximum number of larvae appears to be usually very near to the ground, the pupae tend to be higher up the tree. Irregular distribution is commoner on smooth bark than on well-creviced trees. The amount of alga and lichen controls the distribution of the larvae, but the pupae tend to be more affected by the distribution of cracks in the bark which act as places for pupation."

This unusually thorough documentation of what a particular species requires from the places where it can live shows how subtle these requirements may be. Recapitulating, we note that *Luffia* can live in places where (*a*) the mean temperature during July is not below 62° F. and the temperature during the winter is not low enough to kill the larvae; (*b*) there is an abundance of a particular lichen, but not if the algal component is present alone; (*c*) there is the right amount of light; (*d*) there are suitable crevices for pupation. And, finally, even places having all these attributes have little chance of being occupied unless they happen to be within a mile or two of a population of *Luffia*, for the adult of this species is wingless. The larvae disperse by floating in the air with the aid of silken threads, but McDonogh considered that about 1½ miles was about the greatest distance that they were likely to travel.

Many holometabolous insects spend most of their lives in or on the plant or animal (or debris such as log, carcass, or dung) which serves both as food and as a place in which to live; and it is scarcely practicable with them, during this stage of their life-cycle, to analyze these two components of environment independently. Some species may leave the place where they have been feeding and seek another sort of place when the time comes to pupate, but other species may not. It is interesting that, in species with a facultative diapause intercalated into a multi-voltine life-cycle, the diapausing and nondiapausing individuals usually seek different sorts of places. The former often go farther or search more persistently, with the consequence that they end up in more sheltered places in which they spend the winter, or the summer, as the case may be. This can be readily observed in the codlin moth *Cydia pomonella;* in this species one or two generations in which few individuals enter diapause may be succeeded by a generation in which all enter diapause. Diapause, when present, occurs at the close of the final larval instar. The inception of diapause is largely determined by photoperiod. Those larvae which have experienced long days as they were developing are satisfied to spin a flimsy cocoon under almost the first piece of bark they find, no matter how inadequate the shelter it provides; they pupate at once and emerge as adults without delay. Those larvae which have experienced short days as they were growing show great persistence in pushing or chewing their way into tight cracks or crevices in the bark, where they spin a heavy, dense cocoon in which they spend the winter in diapause.

The beet webworm *Loxostege sticticalis* in Montana has one or two generations during summer, and the winter is spent as a diapausing prepupa. Exceptionally, a small proportion of the second generation may go on to produce a third generation in the one summer (Pepper and Hastings, 1941). The proportion of the first generation which enters diapause may vary from less than 1 per cent to more than 60 per cent. The larvae, on reaching maturity, wander away from the plant and burrow several inches into the soil, constructing a strong silk-lined cell. Those which are going to pupate without delay usually choose loose soil for this purpose, but those which have been determined for diapause tend to wander farther and usually seek out firm soil or sod in which to construct their cells; this is irrespective of what generation they may belong to.

The mite *Bryobia praetiosa*, living on apple trees in South Australia, passes through several generations during summer and then spends the winter as a diapausing egg. All nondiapausing eggs are laid on the backs of the leaves, but all diapausing eggs are laid on the main branches of the tree. It is likely (by analogy with other species) that the mite responds to photoperiod. The striking feature is that exposure to a certain length of day during the early stages of the mite's development determines not only the sort of egg which is to be laid but also the behavior of the mite in seeking a place to lay it. The survival-value of the adaptation is clear, but this makes it no less remarkable.

Fisher and Ford (1947) studied a colony of the moth *Panaxia dominula* (Arctiidae) which occupied about 20 acres of fenlike marsh at Dry Sandford near Oxford. Part of the marsh was wooded, but most of it was covered by reeds and herbaceous plants, including comphrey, *Symphytum officinale*, which was the chief food of the larvae, and several sorts of nettles, which were also suitable for food. The marsh was bounded by woodland and agricultural land, into which the moths seemed never to penetrate, notwithstanding that they flew powerfully, were often observed circling around and above trees in their chosen area, and certainly were quite active in dispersing throughout the 20 acres in which they lived. There was another small area at Tubney, about 1½ miles away, which was suitable for *Panaxia*, but Fisher and Ford found clear evidence that very few, if any, moths found their way from Dry Sandford to Tubney (sec. 5.01). Moreover, entomologists have collected in this vicinity for many years, and their testimony (reliable because *Panaxia* is a large brightly colored day-flying species) confirms that *Panaxia* is not to be found straying beyond the confines of the specialized territory where the colony lives.

This unmitigated adherence to such a small area (in the case of Dry Sandford a mere 20 acres) on the part of a large moth capable of powerful flight indicates a keen response to the boundaries of the area where there are suitable places for it and its larvae to live and an altogether remarkable inhibition of the usual tendency for at least some individuals to disperse widely from the place where they originated (sec. 5.6).

McCabe and Blanchard (1950) made a thorough study of the ecology of three species of deer mice, *Peromyscus maniculatus*, *P. californicus*, and *P. truei*, in an area of about 25 square miles near Berkeley, California; their report is especially interesting for the information they give about the sorts of places where these three species may live. The area studied was an outlying ridge of the Contra Costa hills, with a general elevation of about 1,500 feet and with several peaks approaching 2,000 feet above sea-level. The ridge was dissected by many valleys (canyons), the floors of which might be as much as 1,000 feet below the crests of the adjacent hills. Several different types of vegetation were recognized, and their distributions seemed to indicate that they required different amounts of moisture. The transition from one type to another was often abrupt, and the boundaries were thus often clearly defined. The crests of the moister hills carried woods of *Pinus* or *Eucalyptus*. Grass and *Artemisia* occupied the drier hilltops. The stream beds were lined by narrow bands of moisture-loving trees, *Umbellularia californica*, and oaks also occurred at the lower levels. The hillsides, especially the moister aspects, were characteristically covered by a dense tangled scrub known locally as "chaparral." In different parts of the area the chaparral abutted on all the other vegetation types, and often the transition from chaparral to forest, to grassland, or to woodland was abrupt, giving a well-defined edge to the chaparral. Artificial margins to the chaparral also occurred, perhaps along an old road or a track or some other sort of clearing.

Throughout the whole area the mice were restricted to the margins of the chaparral, whether natural or artificial. Assiduous trapping in the body of the chaparral failed to discover a single mouse; no residents were found in the wooded areas or in the grassland or, indeed, anywhere else except along the very margins of the chaparral. In some places, especially on the more arid slopes, the margins were poorly defined, as the chaparral gradually thinned out and became invaded by grass or *Artemisia;* at the same time, some plants of the chaparral would penetrate the grassland or the area occupied by *Artemisia.* From these and other places where the margins to the chaparral were poorly defined, *Peromyscus* was absent also.

The places most favored by *Peromyscus maniculatus* occurred on the moister slopes, where a luxurious growth of chaparral, composed chiefly of *Bacharis,* gave way abruptly to a narrow band of giant herbs, chiefly Umbelliferae, which, in turn, gave way to grassland. The ground under the *Bacharis* was bare except for a thin covering of dead leaves from *Bacharis.* This obviously suited *P. maniculatus,* which has the habit of seeking shelter in holes and small crevices in the ground and is not able to make very good progress running over a rough surface; but there were doubtless other more subtle qualities which made these the most suitable places for *P. maniculatus* to live. Elsewhere the chaparral manifested clear-cut margins where it met forest or woodland or even a glade with a few trees in it. These margins also provided suitable places for *P. maniculatus;* but they were less likely to be occupied than the other sort of margin, and when they were occupied, the populations were less dense. Especially in the wetter spots, the ground under the chaparral where it joined forest or woodland was covered with a "duff" of dead leaves, accumulations of fallen twigs and branches, and a tangle of low herbage. The presence of the duff on the ground was disadvantageous to *P. maniculatus,* but doubtless there were other ways in which this sort of margin failed to meet the requirements of this species.

The other two species seemed never to seek shelter underground but always in the debris of vegetation on the surface. Consequently, places where the ground was covered with duff suited them quite well. They were most likely to be found and their populations more likely to be dense along the edges of the chaparral, where it met woodland of *Pinus, Eucalyptus,* or *Quercus,* or around the edges of a glade. It was characteristic of the places favored by *Peromyscus truei* and *P. californicus* that the chaparral would be composed of a greater variety of species, its margins would be more sinuous and less linear than those which were favored by *P. maniculatus*, and the ground would be covered by duff. Their requirements were more stringent than those of *P. maniculatus*, for they were always very few except in the places that suited them precisely. On the other hand, *P. maniculatus* was almost ubiquitous, being present, if not abundant, wherever the chaparral manifested a well-defined edge.

In some of the areas best suited to *Peromyscus maniculatus* this species contributed more than 95 per cent of all the mice captured. In the places which were best suited for *P. truei* and *P. californicus*, these two might contribute as much as 75 per cent of the total catch, with *P. maniculatus* making up the rest. The ratio of *P. truei* to *P. californicus* varied from 3:1 to 9:1. No place was found where the population consisted exclusively of one species. The places where the relative numbers of the three species were most nearly equal were also the places where the absolute numbers were lowest.

In a number of experiments the trapped mice were destroyed instead of being released in the same place. This never resulted in any change in the relative proportions of the different species in that area. From these and other observations McCabe and Blanchard concluded that neither competition nor any other sort of interaction between the species was important in determining either the absolute numbers of any species or the relative numbers of the three species in an area. On the other hand, McCabe and Blanchard observed what was at least the rudiments of territorial behavior in these mice, and this, in relation to the number of good places to live, seemed to be what chiefly determined the density of the population of each species. Since any particular area seemed to carry a characteristic number of each species, it may be safely inferred that each species had its own special requirements with respect to the place where it would live.

The spotted skunk *Spilogile interrupta*, like other small vertebrates, requires in its territory one or a number of "dens." Crabb (1948) described a den as "any location or cover which the animal uses of its own free will for rest or seclusion." He studied the sorts of places which were used as dens and concluded: "The first consideration seems to be the exclusion of light. Without exception every den or semblance of a den met this requirement. Sometimes the den was only a place to curl up or stretch out in, such as under the corner of a well platform or a shock of small grain, but wherever darkness prevailed there the spotted skunk seemed to feel most at home." The den also provides shelter from the weather and protection from predators, including men and domestic animals. Crabb considered that the number of places (hollow logs, burrows, spaces under woodpiles, and so on) which would meet these requirements around the farms and fields of southern Iowa were almost unlimited. But, of course, some would be very much better than others.

In certain months of the year the number of rats in residential blocks in Baltimore was found to be proportional to the amount of broken paving, which is a measure of shelter available. At other times shelter was not limiting, but food or some other component of environment was (Davis, 1953).

Leopold (1933) described the qualities required in a "deer range" in the Lake states. The deer require a cedar swamp for "yarding" during deep snow; for hiding and sleeping, they need evergreen thickets, preferably on the point of a saddle on a hill; for play, open places are needed; and for fawning, the doe must be near water where she can satisfy her thirst during nursing without having to travel too far. Leopold made the generalization that "a range is habitable for a given species when it furnishes places suitable for it to feed, hide, rest, sleep, play, and breed all within reach of its cruising radius." Birds often include in their requirements not only a suitable site for a nest but also places for singing, for roosting, and, in resident species, a secure place for sleeping during winter. The crossbill and the tree pipit require tall trees for singing posts; the nightingale and the wren sing from vantage points much nearer to the ground (Lack and Venables, 1939). The special requirements of the bobwhite quail and the muskrat have been studied quite thoroughly, but we shall leave the discussion of these two species until section 12.3.

12.11   *Specialized Behavior Associated with the Choice of a Place To Lay Eggs*

With insects it is usually, though not always, the adult which is specialized for dispersal, and it is in this stage that are found the most remarkable adaptations for seeking out the right sorts of places to deposit eggs. In striking contrast to *Panaxia*, which remains so firmly attached to its home territory (sec. 12.1), the two butterflies *Danaida archippus* and *Pieris rapae* are remarkable for their widely ranging flights in search of plants on which to oviposit. The larvae of *Danaida* live and feed on the cotton bushes (*Asclepias* spp.). In southern Australia these shrubs grow on roadsides, in parks, and on wastelands; they are distributed widely but often quite sparsely. Yet it is unusual to find one during summer that does not carry at least a few larvae of *Danaida*. Similarly, the larvae of *P. rapae* are found only on plants of the family Cruciferae; and one is impressed by the high probability that even remotely situated plants of this family will be found and oviposited on by this widely ranging butterfly (see also sec. 11.34).

The mosquito *Anopheles culifacies* lays its eggs in rice fields when the rice is not very high; when the rice plants reach a height of about 1 foot, the mosquitos disappear. The female while laying eggs flies a tortuous course just above the surface of the water. If it is prevented by the presence of tall rice plants from doing this, then it will not lay eggs in that place. An otherwise suitable place can be made quite unsuitable by simply inserting all over it glass rods which project about a foot above the surface (Macan and Worthington, 1951). There is doubtless survival-value in this odd behavior; it would be interesting to know whether this habit prevents the mosquito from laying eggs in places where the larvae would have little chance to survive. In North America there are 8 species of mosquitos whose larvae are found only in rain-filled rot-holes in trees (Jenkins and Carpenter, 1946). There are 3 species in Britain similarly restricted to tree-holes (Elton, 1949). It has been suggested that the gravid females of these species are attracted to the holes and stimulated to lay their eggs in them by the presence of some organic substance which is absent from water accumulated in other sorts of places. But this is not known; nor is it known what other qualities make the tree-hole a suitable place for these species to breed in.

The larva of the beetle *Lyctus brunneus* lives in freshly seasoned timber, provided that it contains a sufficiency of starch and not less than 8 per cent of moisture. Starch is absent from the true wood of all sorts of trees, but it is present, often in adequate concentrations, in the sapwood of certain broad-leaved species. The beetles seek a place where the vessels of the xylem (which run in the sapwood) have been cut across or otherwise exposed, and place their eggs inside the cavities of the vessels. Apart from rare accidents, the eggs are never laid anywhere else. The eggs of *Lyctus brunneus* are about 180 μ wide. The beetles usually choose vessels about this size; they will not put eggs in wide vessels where they would be a loose fit, and, of course, they cannot put them in vessels that are narrower than the egg. Consequently, the timber of coniferous trees and the true wood of broad-leaved species do not have any eggs laid in them, nor does the sapwood from broad-leaved trees, in which the vessels of the xylem are too narrow. The first two would not contain enough starch to support the larvae, but the last may; in that case its unsuitability as a place for *Lyctus* to live depends entirely on the behavior of the beetles in refraining from ovipositing anywhere except in vessels of such a size that the ovipositor can just be neatly pushed into them (Parkin, 1934; Gay, 1953).

Frogs of the species *Rana temporaria* in England emerge from hibernation some weeks before spawning begins. They frequently leave the pond where they hibernated and travel, often quite a considerable distance, to another pond to spawn. They usually place their eggs in a few restricted areas within the pond. The whole phenomenon is rather puzzling because often the abandoned pond looks quite suitable, and the new pond often appears far more uniform than the patchy distribution of eggs would indicate: just what it is that makes a place suitable for *Rana* to lay its eggs remains a mystery (Savage, 1934, 1935).

Birds usually have specific requirements for nesting sites and nesting materials. In Britain certain species of the Corvidae and Falconidae nest only in the canopies of tall trees and may be found only in woods which contain at least one tall tree; each species is characterized by a minimal height below which it will not build. But the nightingale and the wren tend to keep in the secondary growth, well below the canopy, for all their activities during the breeding season. Many species of tits, woodpeckers, and other birds of the woodland require holes in trees for their nests. That is why they are often scarce or absent from young woods and more common where there are old trees. The scarcity of jackdaws and starlings in woods composed entirely of coniferous trees is probably due to the absence of good holes for nesting in these woods. The absence of the nightingale from coniferous forests has a different explanation: it requires the dead leaves from deciduous trees for its nest (Lack and Venables, 1939). In California and Arizona there are two species of woodpecker which make nests by carving a hole in the large upstanding stem of the giant cactus *Cereus giganteus*. In the absence of the cactus, they will also make nests in trees. The elf owl *Microphallus whitneyi* nests exclusively in the abandoned nests of the woodpeckers in the cactus, never in any other tree. This is often quoted as an example of the dependence of one species on another (Elton, 1927, p. 48; Allee *et al.*, 1949, p. 362), but it would seem that none has described the subtle difference that the owl discerns between an abandoned woodpecker's nest in *Cereus* and one in some other kind of tree.

12.12   *Some Clues from Studies of "Succession"*

Studies of the "ecological succession" which may go on in such habitats as a fallen log, a carcass, or a heap of dung belong properly to community ecology. Unfortunately for our purpose, they rarely indicate, except in very broad terms, what any one species requires. We mention several such studies which do indicate how the different species in the succession may have quite different requirements of the places in which they live.

Savely (1939) studied the animals that lived in rotting logs of pine and oak in a forest in North Carolina during successive stages in the logs' decomposition. The first species to enter the fallen pine logs were 22 species of beetles, chiefly Cerambycidae and Scolytidae; these lived and fed almost exclusively in the phloem. The bark soon became loosened from the wood, and the space between wood and bark became packed with chewed-up wood. This favored the growth of fungi. Once the fungi had become established, the log became habitable for a further 37 species of insects and mites which fed on fungi and decayed wood. The greater number of species (29 out of 37) were Coleoptera, but the species having the greatest numbers of individuals were Collembola, Acarina, and Diptera. By the second year the phloem-feeding species had disappeared, the wood was becoming soft and was invaded by 11 new species, among which the termite *Reticulitermes flavipes* was prominent. After this, no marked changes occurred in the occupants of the log as it gradually became incorporated into the soil.

Mohr (1943) made a study of "succession" among animals living in cattle droppings at Urbana, Illinois. Altogether, 150 species were found in the dung at one stage or another from when it was fresh to when it finally became incorporated into the soil. The particular insects breeding in the dung at any stage were largely determined by the microörganisms which decomposed it; but the animals also played a part in determining the "succession." The maggots of *Cryptolucilia* and *Sarcophaga* produced galleries and openings in the dung which were later used by certain staphylinid beetles. It seemed that the beetles depended upon the galleries made by the previous inhabitants, for they used them continuously without making any new ones for themselves.

Woodroffe and Southgate (1950) described the animals which lived in the nests of sparrows in England. At first, while the nest is being more or less permanently occupied by the birds, the other inhabitants are nearly all ecto-parasitic arthropods, insects, ticks, and mites. When the nest has been abandoned by the birds, it is invaded by scavengers; these are largely moths of the families Tineidae and Oecophoridae, beetles of the families Ptinidae and Dermestidae, certain silverfish, and mites. Many of the species are the same as those found in warehouses and stores, feeding on grain and other sorts of human foodstuffs. After a period the nest decomposes into a humus-like mass, and the foregoing species are supplanted by another group, largely composed of the sorts of animals that are usually found in soil. A similar sequence may be observed in a carcass (Fuller, 1934; Waterhouse, 1947; see also sec. 10.31). Some examples of succession of animals associated with plant succession are summarized by Odum (1953, pp. 190–94).

12.2   SOME EXAMPLES OF THE INTERDEPENDENCE OF "PLACE TO LIVE" WITH OTHER COMPONENTS OF ENVIRONMENT

12.21   *Temperature*

For many species of terrestrial invertebrates which live in the cool, temperate regions of the Northern Hemisphere the size of the population which resumes activity in the spring may depend largely on the number which has been able to survive exposure to cold during the winter. Special adaptations, such as diapause and the cold-hardiness associated with it and the tendency to seek well-protected situations in which to spend the winter, are important. But some individuals find better winter quarters than others, and the survival-rate may depend partly on the severity of the winter and partly on the proportion of the population which happens to find adequate quarters. Salt (1950, p. 285) expressed this principle well: "Low winter temperatures are a limiting factor in the survival of many species of insects and often restrict their geographical range. Although the few species which hibernate in very exposed situations can survive the formation of ice in their tissues, the remainder are killed if they become frozen. The latter depend for their survival on the insulating protection of their hibernacula (soil, plant debris, snow, ice, etc.) and on their ability to undercool. The combination of these two protective factors may be sufficient that mortality as a result of freezing is a rarity in some species. More often the protection is incomplete, with the result that those individuals in the more exposed hibernacula and with the least ability to undercool perish. In severe winters this fraction may be temporarily large and may seriously deplete the population. If a species has been expanding its range into colder regions, one severe cold period can eradicate it in such areas, restricting it once again to its normal range."

Mail (1930, 1932) and Mail and Salt (1933) measured the temperature of the soil at several depths in Minnesota and Montana and related this to the probable survival-rate of certain insects which spend the winter in the soil.

From experiments in the laboratory with cold-hardy adults of the Colorado potato beetle *Leptinotarsa decemlineata*, Mail and Salt (1933) concluded that even a brief exposure to $-12°$ C. would kill virtually all the beetles; that more than half of them would die if they were exposed for some hours to $-7°$ C.; and that even prolonged exposure of days or weeks to $-4°$ C. would kill very few beetles. From their own experiments and the observations of others, they concluded that the depth to which the beetles burrowed into the soil at the onset of winter depended on the soil; in loose, sandy soil the majority would be between 14 and 24 inches, but in harder soils they might all remain in the top 8 inches, concentrated just above the "plow line."

The temperature experienced by the beetles during winter depends not only on their depth below the surface of the soil and the prevailing temperature of the air but also on the amount and permanence of snow lying on the ground during the cold weather. Snow is a good insulator, and the presence of 6 inches of snow on the ground results in quite moderate temperatures in the soil below it (Figs. 12.01 and 12.02). The data for Figure 12.01 (*lower curve, with snow*) were collected at Bozeman, Montana, on February 9, 1932. Although it was an unusually cold day, beetles that were hibernating more than 12 inches below the surface would have been quite safe. Most of those which had remained in the top 12 inches, either fortuitously or because they had been prevented by hard soil from burrowing more deeply, would have been in danger during this cold spell. The data for Figure 12.01 (*upper curve, no snow*) were collected at St. Paul, Minnesota, on January 1, 1928, in soil from which the snow had been scraped away; the weather was not unusually cold. In the absence of snow, the soil was much colder, and only those few individuals which happened to have gone deeper than 2 feet would have had much chance to live through the winter.

In the regions where Mail and Salt worked, the occurrence of warm "chinook" winds may cause the snow to disappear quite suddenly even in midwinter. Then the temperature of the soil down to a depth of several feet may fall quite dramatically. Figure 12.02 illustrates one such occurrence; beetles hibernating in the top 2 feet of soil would have had little chance of surviving this cold spell. On the other hand, Mail (1932) observed that during the winter of 1931–32 snow remained almost continuously from December 1 to March 31 on the area where he had his instruments, and the temperature of the soil even near the surface fell no more than a few degrees below zero. At no depth was it cold enough to kill many of the overwintering adults of *Leptinotarsa*.

This species is permanently established in Montana, and it must be concluded that during the more severe winters the population is carried forward because at least some individuals happen to find winter quarters in specially

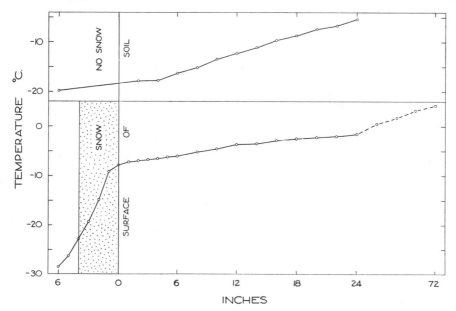

Fig. 12.01.—Temperature of the soil at various depths below the surface: *upper curve,* in the absence of snow; *lower curve,* when snow was present. Note the steep gradient through the 4 inches of snow and the consequent less severe temperature in the soil below the snow. (After Mail, 1930.)

favored situations. The beetles may survive near the surface in places where the snow is more permanent; in more exposed places they have little chance of surviving unless they are buried deeply in the soil. The beetle has not extended its distribution northward into Alberta beyond about the 54th parallel, despite the presence of plenty of food there. This is probably because in this region the temperature of the top 2 feet of soil is likely to fall below −7° C. at least once during most winters (Mail and Salt, 1933).

The interaction between temperature and place to live is nicely illustrated by these studies. The important qualities of the place where *Leptinotarsa* may spend the winter are the nature of the soil, which may determine how deeply the beetles will burrow, and the amount and permanence of the covering of snow. At Bozeman (latitude 45°) *Leptinotarsa* may have a good chance of surviving the winter in situations where the snow is permanent during the winter; in other places, only those individuals which happen to be buried more than 2 feet below the surface have much chance of surviving the winter. At Beaver-lodge (latitude 55°) the prevailing atmospheric temperature is lower, and even the most sheltered place may be too cold for *Leptinotarsa*.

During 1924 the noctuid *Euxoa segetum* was abundant in the vicinity of the Lower Volga, doing much damage to crops. It is unusual for this species to be so abundant in this area, though outbreaks occur rather frequently in adjoining districts. Sacharov (1930), in seeking an explanation, explored several hy-

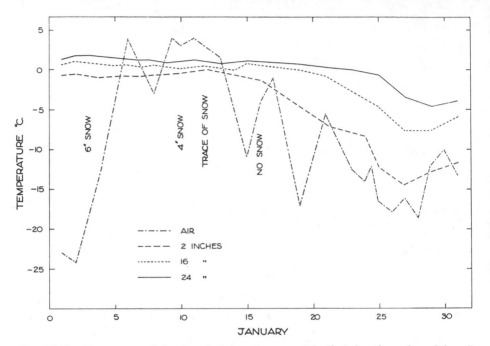

FIG. 12.02.—Temperature of the air and of the soil at several depths below the surface of the soil. Note how, at first, the presence of snow kept the soil warm during a very cold spell of weather and how, after the snow had disappeared, the temperature of the soil fell. (After Mail, 1930.)

potheses relating to food, predators, and the weather during summer but found that none was consistent with the facts. He then examined the survival-rate of overwintering larvae in relation to records of atmospheric temperature, the amount and permanence of snow, and the depth to which the larvae burrowed into the soil; and he found a nice explanation in terms of the interaction between temperature and the places where the larvae spent the winter. There are two generations of *Euxoa* during summer. The larvae of the second generation, maturing during autumn, accumulate large reserves of fat, and their tissues contain relatively little water. Such larvae burrow into the soil to a depth of 6 or 8 inches, often well before winter. But other more tardy individuals of the second generation require to continue feeding later, and these may be trapped near the surface and killed by exposure to cold during winter. Even the others which are relatively cold-hardy because they are in diapause (sec. 6.314) and relatively well protected because they are deeper in the soil may be killed by exposure to cold if the soil does not remain covered by an adequate depth of snow. In one experiment Sacharov scraped the snow away, so that the surface of the soil remained bare, and found that 77 per cent of the diapausing larvae which were overwintering about 7 inches below the surface were killed by the cold.

In the laboratory Sacharov found that diapausing larvae were not harmed by 18 hours' exposure to $-6°$ C. but were killed by exposure to temperatures below $-8°$ C. He related these findings about the behavior and physiology of *Euxoa* to his observations of their distribution and abundance. His conclusions are best given in his own words:

> If we study, further, the January isotherms of the region, we will see that the regions most affected by the insect are the coldest, but at the same time they have the deepest snow cover, which serves to protect the hibernating caterpillars from frosts. In the south-eastern steppe and semi-desert districts winter temperatures are higher than in the forest-steppe zone; but the very thin snow-cover and continual dry east winds cause freezing of the soil to a greater depth, resulting in the killing of caterpillars by autumn and winter frosts.
>
> An example of the fluctuations of soil temperature in connexion with the snow-cover is presented by the observations made in Saratov district for the last three years. During the winter of 1923–24 the snow cover was 12.6 cms. in December, 19 cms. in January, 29.6 cms. in February, and the soil temperature did not drop below $-5.5°$ C. During this winter the deep snow cover and the moderate freezing of the soil were favorable to the hibernation of the caterpillars, and, as a result, very heavy infestation and damage were recorded in the summer of 1924. In the next winter, 1924–25, the snow cover during December, January and February did not exceed 4.5 cms. for each of these months, and the soil temperature at a depth of 25 cms. dropped in December to $-12°$ C., that is, to a level at which the majority of the caterpillars, according to our laboratory experiments, should die. Indeed the damage recorded in 1925 was very slight. Since no parasites or diseases of caterpillars have been observed, this reduction in numbers can be ascribed only to the deep freezing of the soil in December. Further, the winter of 1925–26 was similar to that of 1923–24, as far as the soil temperature was concerned, but the snow cover was as in the winter of 1924–25. One might have expected an outbreak of *Euxoa* next summer, but this did not happen, simply because more than one year is necessary for the insect to increase in numbers to the extent of becoming a pest.

The temperature inside plants or on the surface of leaves, bark, etc., where small animals may live may be quite different from that of the surrounding air. The temperature of one leaf may vary quite markedly from that of another, depending on their exposure to radiation and wind (Wellington, 1950). All stages of the beetle *Dendroctonus brevicornis* live inside "galleries" which they cut into the bark of pine trees. Miller (1931) inserted thermometers into the bark of a tree on its southern and northern aspects. The bark on the south side was warmer and that on the north side cooler than the air in the shade, and the difference between the two was about $8°$ C. The insects which happened to be living on the south side might be expected to develop more rapidly than those on the north side. Henson and Shepherd (1952) inserted thermocouple junctions into the mines made by the caterpillars of *Recurvaria milleri* in the needles of lodge pine. During the day when the sun was shining, the temperature inside the leaf was as much as 7° C. higher than that of the air in the shade. There was a close linear relationship between the amount of radiation falling on the leaf and the difference between the temperature of the leaf and the air. This depended not only on the extent to which the leaf was shaded but also on the angle made by the leaf with the sun's rays. The temperature of leaves in the most shady situations differed little from that of the air in the shade, so there was a difference of as much as 7° C. in the temperature experienced by different individuals of *Recurvaria*, depending on whether they happened to be living in leaves that were well shaded or exposed to much radiation.

Uvarov (1931) suggested that bubonic plague survived in the steppes around the Caspian Sea because the fleas on the ground squirrel *Spermophilus* were protected against the cold of winter and the dry heat of summer in the burrows of the squirrel. Similarly, Buxton (1932b) found that rat holes in Palestine provided a suitably cool and moist place in which *Xenopsylla cheopis* could live, although the temperature and humidity recorded at midday in meteorological screens near by would have been fatal, at least to the larvae.

12.22 *Moisture*

In warm, temperate zones the hot, dry summer may be the most hazardous season for small terrestrial animals; and we can recognize, in the species from these regions, adaptations which lessen the dangers from desiccation. Diapause is important (sec. 4.6), and so is behavior in seeking the most favorable places to aestivate. Still, chance plays a large part, and an unusually severe summer may kill all but the few individuals which happen to be in the best places. This happened to the grasshopper *Austroicetes cruciata* in parts of South Australia during the summer of 1938–39.

This grasshopper aestivates as a diapausing egg. The females can, but in nature they rarely do, lay their eggs in soil that is loose or soft. Instead, they invariably choose soil that is hard and compact; bare, windswept patches are favored, and they have even been observed boring into the edge of a macadam road. The female seems to experience difficulty in starting the hole unless she can keep the tip of her abdomen pressed firmly against the surface of the soil. A female may often be seen to back up to a suitable obstacle, such as the side of a stone or a small bush, flex her posterior legs upward, grip the stone about half an inch from the surface of the ground, and use this leverage to press the tip of the abdomen firmly against the surface of the soil while starting to bore. The result is that many eggs are laid around the edges of such obstacles, though many are also laid out in the open; minute cracks and other minor irregularities in the soil may be exploited in the absence of stones or small bushes.

During the summer of 1938–39 some districts where *Austroicetes* occurs went for 89 days with virtually no rain; this was an extraordinary drought. In one district where it was estimated that at least 90 per cent of the eggs had died, none was found alive except among those that had been laid around the edges of large, flat stones. Many had died even in these situations, and it was specially noticeable that, with very few exceptions, the eggs in any one pod (usually about 20) were either all dead or nearly all alive (Andrewartha, 1939; and unpublished notes). The inference is quite plain. The only eggs which had any chance of surviving this drought were those which had been laid around the edges of stones large enough to provide "runoff" from the few scant showers of rain that occurred; even so, only those few survived which, by virtue of local vagaries in the contours of the stone or the soil, received rather more than their share of the small amount of water thus provided. Farther south, where the drought was less severe, the survival-rate varied from 36 to 78 per cent, but it was still noticeable that small local differences in topography largely determined which eggs should survive and how many.

The distribution of the cricket *Gryllulus commodus* in South Australia is strikingly limited by the distribution of the type of soil known as "rendzina." In this region these soils mostly occur in long narrow strips, because they form the floors of narrow valleys separated by low, sandy hillocks which were once coastal sand dunes. These soils support permanent populations of *Gryllulus*, because, with the weather experienced in this region, the eggs have little chance of surviving unless they are laid in a soil with a very high water-holding capacity. The full explanation for this was worked out by Browning (1952, unpublished thesis).

With *Gryllulus* the life-cycle occupies one year. The eggs are laid during April or May (southern autumn); diapause-development is completed early in winter, but the eggs do not hatch until early summer, usually during November. In some years a high proportion of the eggs give rise to nymphs, but more often only a few survive in the most favorable situations. Most of the deaths occur during spring, when the eggs are nearly ready to hatch. The chief cause of death is desiccation due to the drying-out of the soil around the eggs.

Compared to hardy eggs such as those of *Austroicetes cruciata* (sec. 7.233), the eggs of *Gryllulus* are poorly waterproofed. Browning found that none hatched unless they were kept in contact with free water or in an atmosphere saturated with water vapor; in one experiment 1 day's exposure to a relative humidity of 90 per cent at 29° C. killed 75 per cent of the eggs. In nature the eggs are laid about ¼ inch below the surface of the soil, and Browning's experiments show that an egg is unlikely to survive if the soil around it dries out to the "wilting point" even for 1 day.

The climate in this region is of the Mediterranean type; that is, the winter is mild and humid and the summer is warm and dry. The soil remains continuously wet during winter but dries out, at least on the surface, for several months during summer. The chief hazard for the eggs is that the surface layer of soil in which they occur may become dry before they hatch. The risk is less in soils of high water-holding capacity. The rendzinas are heavy black soils, very sticky when wet, with a high proportion of clay and organic matter near the surface; they have a high water-holding capacity.

Occasionally the crickets become very numerous and do a lot of damage to pastures; this is likely to happen after 1 or 2 years in which the rainfall during winter was high, especially if it lasted well into the spring. During the drier, more normal years, when most of the eggs die without hatching, the survival-rate varies widely from place to place. This is associated with variations in the quality of the soil. The rendzinas in this region are highly variable, especially with respect to the important quality of water-holding capacity, which is associated with heaviness and blackness. In the more severe years the higher survival-rates are associated with the heavier and blacker patches of rendzina. In the kinder years the survival-rate may also be high on the lighter variations of this soil type, though scarcely ever on any of the other soil types which occur in this region.

Although the association of *Gryllulus* with the rendzinas is close and almost general, the few exceptions are important because they confirm the theory. Browning found one area of several acres in which the crickets were thriving on a light soil (meadow podsol); the local topography was such that a high water table kept the surface soil in this area moist even in the absence of rain. Elsewhere *Gryllulus* occurs in numbers on almost any sort of soil in gardens around dwellings where the soil may be watered during summer.

In much the same region where *Gryllulus* is found, the distribution of the moth *Oncopera fasciculata* is virtually restricted to the lighter soil types, scarcely ever being found on the rendzinas. The soils on which *Oncopera* is found may be broadly grouped into volcanic soils and meadow podsols. The former are mostly deep, friable, and well drained; they may dry out rather severely during summer but never become excessively wet, even during the wettest of winters. The latter are mostly found in lower situations; the subsoil is heavy and the surface soil may become water-logged during winter.

The life-cycle of *Oncopera* occupies 1 year. The moths are present for a few weeks early in spring; the eggs hatch during October or November, and the larvae are present from then until late in the winter (July–August). The larvae establish themselves in vertical burrows in the soil, from which they emerge during the night to feed. They remain relatively dormant during the summer, which is dry, and feed and grow most actively during the winter, which is mild. A large proportion, amounting at times to virtually the whole population of an extensive area, may die during the summer from excessive dryness or from drowning during winter. The risk of desiccation is considerable for the larvae in both the volcanic soils and the meadow podsols but is greater for the former. The risk of drowning is negligible for larvae on volcanic soils but is quite great for those on meadow podsols. Notwithstanding these twofold risks, exceptional weather may allow a high survival-rate on either sort of soil, and, if this recurs for several consecutive years, dense populations may be built up over quite extensive areas. The multiplication and decline of this species are determined largely by the survival-rate during the larval stages, and this depends mainly on the interaction of weather (chiefly moisture) and the places where the larvae may live (Madge, 1953, unpublished thesis).

In their pristine condition the plains in the region where *Oncopera* is found carried certain perennial grasses with a tussocky habit of growth. The larvae of *Oncopera* tend to extend their burrows upward into the crowns of these tussocks, a habit which enables them to escape drowning in the wetter situations. With the development of the land for agriculture, these grasses have disappeared, and the *Oncopera* living in the modern pastures have no opportunity to practice this habit. But Madge came across one farm on which, because of certain accidental circumstances in the establishment and subsequent management of the pastures, there was a meadow in which a dense low growth of clover was studded, at intervals of several feet, with tussocks of cocksfoot, with crowns rising some 6 or 9 inches above the level of the ground. It also happened that this meadow carried a dense population of *Oncopera*. During the autumn the larvae were present at the rate of four or five per square foot in the flat parts, and each tussock also contained four or five larvae; their burrows extended down among the roots and up into the crowns (Fig. 12.03). Heavy rain fell during May and June, and the soil became water-logged (Fig. 12.04). So far as could be ascertained, all the larvae whose tunnels were in the flat ground between the tussocks were drowned, but hardly any of those which happened to be living in the tussocks were killed.

In northern England the life-cycle of the sheep tick *Ixodes ricinus* occupies 3 years. About 6 months after emerging from the egg, the larva goes in search of its first meal in the spring; it climbs to the tip of a grass stem and awaits the passing of a sheep or some other small animal. If it has the good fortune to be picked up by a host, it engorges itself with blood and after a few days drops to the ground. Being incapable of walking more than a few inches, it is unable to seek for a good place to live; but if it is fortunate enough to have dropped into one where the shelter is adequate, it may survive the 12 months which must elapse before it is ready to look for its next meal. During this time it molts and grows into a nymph. At the appropriate season (April–May) the nymph again climbs a grass stem and awaits the passing of a sheep. The nymph, after engorging with blood, drops to the ground. If it survives during the next 12 months, it becomes adult. In the next spring the adult climbs to the tip of a grass stem, seeking not only a meal of blood but also a mate on the sheep (sec. 9.13). Being fully engorged and having mated, the female drops from the host and burrows, like the larva and nymph, into the vegetation. If she survives, she lays her eggs about 2 months later (MacLeod, 1934; Milne, 1952).

During the three years of its life the tick spends, altogether, about 3 weeks feeding and about an equivalent period on the tips of grass stems waiting to be

FIG. 12.03.—Two sorts of places which may be occupied by larvae of *Oncopera fasciculata* in a meadow containing tussocks of cocksfoot. *Inset:* diagrammatic section to show the larvae in their burrows. *A*, in the absence of flooding, larvae may survive in both sorts of places, *B*, when the soil becomes water logged, the larvae whose burrows are in the open may drown, whereas those whose burrows ascend into the tussocks may escape drowning. (After Madge, unpublished thesis.)

picked up; the rest of the time is spent sheltering amid the vegetation close to the ground. Many ticks must die in each generation from starvation; predators, chiefly shrews and birds, take a heavy toll; but more fundamental than either of these hazards is the need to conserve the supply of water in the body during the long intervals between meals (Milne, 1950*a*, *b*). In its younger stages *Ixodes*, in a starved condition, may absorb water from the surrounding air if the relative humidity exceeds 88 per cent; in its later stages a tick requires more humid surroundings, and an unfed adult cannot absorb water from air that is drier than 92 per cent relative humidity (sec. 7.213). Consequently, the ticks after dropping from the host have little chance of surviving until the next meal is due (or to lay eggs, as the case may be) unless they happen to find a place in which the air is likely to remain saturated with water vapor, or nearly so, for most of the time.

According to Milne (1944, 1946), the distribution of *Ixodes* in northern

Fig. 12.04.—Part of the meadow illustrated in Fig. 12.03 during a spell of wet weather. The soil was water-logged, and all the larvae whose burrows were in the open places between the tussocks were drowned. (After Madge, unpublished thesis.)

England is strictly limited to rough pastures on hills and moors. The vegetation in these pastures is coarse, and there is a dense, almost permanently moist, mat of semidecayed vegetable matter, moss, and grass, covering the surface of the soil to a depth of several inches. In this material the relative humidity is close to 100 per cent for most of the year, and the ticks find adequate shelter from desiccation. Milne (1950a) did a number of experiments to find out the precise location of the ticks in this mat (Table 12.04). On a number of different occasions he placed about 40 newly molted adults on the surface of three different sorts of turf and allowed them to crawl down into it. Once they had settled down, they did not change position throughout the season, and the data in Table 12.04 summarized their distribution in the mat during the rest of the summer and the winter. Most of them penetrated well into the mat, where they were likely to be surrounded perpetually by saturated air. Not all parts of the pasture were equally favorable in this way, and Milne (1950b)

TABLE 12.04*

VERTICAL DISTRIBUTION OF *Ixodes* IN VEGETATION ON ROUGH UPLAND PASTURES
AS INDICATED BY RELATIVE NUMBERS FOUND AT VARIOUS DEPTHS IN VEGETATION

TYPE OF PASTURE	LAYER	NO. OF TICKS AS PER CENT OF TOTAL	
		Males	Females
Rough grass......... (mat 2½ inches)	Grass	3.9	4.8
	Upper mat (top 1½ inches)	84.4	81.2
	Lower mat	10.7	14.0
	Soil	1.0	0.0
Heather............. (mat 2½ inches)	Heather	0.0	0.0
	Upper mat (top 1½ inches)	96.4	82.3
	Lower mat	3.6	15.2
	Soil	0.0	2.5
Bracken............. (mat 1 inch)	Bracken	0.0	0.0
	Mat	100.0	98.7
	Soil	0.0	1.3

* Data from Milne (1950a).

showed that the abundance of ticks was correlated with the amount of cover. In the short swards, which are more characteristic of the lowland pastures, the vegetation offers no place where the moisture is adequate to support the tick throughout the year. Consequently, *Ixodes* never becomes established on the lowlands, despite the repeated introduction of flocks of infested sheep into these areas.

In Scotland the biting midge *Culicoides impunctatus* is known to have a "patchy" distribution. Kettle (1951) not only demonstrated the patchiness

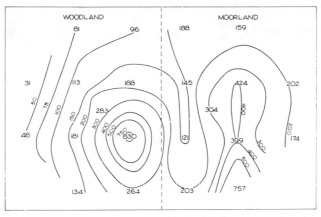

FIG. 12.05.—The distribution of adult females of *Culicoides impunctatus* in a 5-acre field of woodland and moorland at Bannachra. For further explanation see text. (After Kettle, 1951.)

quantitatively but also provided an explanation for it in terms of the sorts of places in which the larvae can live. He placed 20 traps uniformly over an area of about 5 acres and recorded the number of *Culicoides* caught each week for 18 weeks. Half the area was moorland, and half was woodland. The numbers caught in each trap were expressed as a percentage of the total for the area, and results were plotted on a plan of the area (Fig. 12.05). Lines were drawn joining places where the midges were trapped in equal numbers, and these indicated two distinct centers of concentration, from which it was postulated that all the midges in the area were coming from two distinct, probably quite small, breeding grounds. This was proved for the woodland area by taking samples of soil and counting the larvae in them. The breeding ground turned out to be a small area of rather boggy soil carrying *Sphagnum* and *Juncus*. It was flanked by higher ground, supporting bracken or woodland with undergrowth of various shrubs. No larvae were found in the higher ground; apparently the breeding of *Culicoides* is restricted to places where the soil remains moist. The distribution of the adults is limited by the distribution of suitable breeding grounds and the distance they may fly away from the place where they were bred. Kettle estimated this to be about 80 yards.

12.23 *Other Animals*

In many of the examples used in chapters 9 and 10 to illustrate the influence of other organisms on an animal's chance to survive and multiply, it was necessary, in order to keep the discussion realistic, to take into account the sort of place where the particular animal was living. (Indeed, one of the chief criticisms to be leveled against the mathematical models of predator and prey is that they imply that all individuals of the prey may be found equally readily by the predator. This assumption makes these models quite unrealistic.) This emphasizes the close interdependence between "place to live" and "other organisms." It also makes it unnecessary to say very much in the present section (since the matter has already been largely covered in these earlier chapters), and we shall be content merely to describe briefly several examples, selected because they are especially apt. Most of them have been chosen to illustrate how the suitability of a place in which to live may depend on the protection which the place provides from predators.

Udvardy (1951) quoted a description from Turček (1949) of how, in the forests of the North Carpathians, the birds nesting in the canopies of trees were relatively safe from predators except when the trees were defoliated by the feeding of numerous caterpillars of the moth *Lipara dispar*. In every case which he observed, predators killed the birds which were nesting in the trees that were defoliated.

In central North America the chief predator of the muskrat *Ondatra zibethicus* is the mink *Mustela vison*. A muskrat which is well established (that is, it is living peacefully, free from serious internecine strife) in a burrow well situated with respect to surrounding water may be unlikely to fall a victim to a mink. But one which is homeless or poorly situated with respect to the place where it is living has small chance of surviving. Water, preferably sufficient to cover the entrance to the burrow, is necessary for security, since drought is one of the most serious hazards experienced by muskrats in Iowa. A quotation from

Errington (1939) illustrates the way in which moisture may influence the musk-rat's chance to survive in the presence of predators and how, when droughts occur, some individuals have a better chance of surviving than others, because they happen to be living in places which offer greater security than those occupied by less fortunate individuals:

The effect of drought on vulnerability of muskrats to mink is so pronounced nevertheless that a few examples should be appropriate in this paper. For both summers of 1936 and 1937 there were on Round Lake close to one adult muskrat per 2.7 acres and in both years the spring mortality from mink was conspicuous. In 1936, the mink pressure slackened in May but was resumed in July as the water level of the marsh went down. Exposure of the bank burrows along about 300 yards of the southeast shore was followed by the killing by mink, largely between July 22 and August 1, of apparently all but one of eight muskrats believed to be resident there. The one individual known to have escaped was living in a newly built lodge about 60 yards from shore. Similar mortality in other exposed shore habitats was also detected but nearly the entire population of Round Lake was already living in much greater security in lodges deeper in the marsh. In contrast, the summer of 1937 was a season of high water in northwest Iowa, and examination of mink prey items and 168 faecal passages gathered from Round Lake did not disclose any evidence of mink pressure upon muskrats from May to early October. The vulnerability of the muskrats of the Dewey's Pasture potholes, which went dry from June to August, 1936, seemed to become critical just after the disappearance of the surface water. Mink faeces deposited before this time rarely contained muskrat remains; then, with the exposure of the muskrats, mink diet ran strongly to this item for a week or two.

The red fox *Vulpes regalis* rarely preys upon muskrats that are protected by the presence of water around their burrows; but when drought lowers the level of water and exposes the muskrats, the fox may, by virtue of its superior skill, exceed even the mink in the severity of its preying. A quotation from Scott (1947) illustrates these points: "The muskrats lived in a marsh habitat[1] at Wall Lake and in a river habitat[1] at Moingona. Field observations indicated that the foxes tended to avoid wading through water, and muskrats appeared to be secure while in water. There was very little evidence of muskrats feeding on land within the area at Moingona, possibly because there was very little cultivated land adjacent to the river. The remains of muskrat were detected but twice in the fecal material (from foxes) from the Moingona area. At Wall Lake the muskrats were left exposed on land in the 34-acre occupied area when the water levels were lowered by drought during June, July and August of 1940. The unusually severe and apparently somewhat uncompensated predation upon these muskrats has been described by Errington and Scott (1945)." Errington (1943) considered that the foxes killed more muskrats than would have been destroyed by mink in the same circumstances. He estimated that the presence of this family of foxes in the area (which was somewhat fortuitous) had resulted in the deaths of some 75 muskrats which might have survived in the absence of the foxes.

1. In this sentence "habitat" seems to refer not to actual space occupied by the animals but to the sort of country in which they were living, e.g., marshes or rivers in general.

The next example, of the scale insect *Saissetia oleae*, is of special interest because the provision in a district of suitable places protecting it from its predator may result not in an increase in the numbers of the scale insect but in a decrease. When the scale insect *S. oleae* is living in a place where there is a mild and humid climate, it will breed continuously throughout the year; generations overlap, and the population consists of individuals in all stages of development. In places where the summer is hot and arid, breeding may be restricted to the winter. For example, in groves of citrus in the Piru district of Ventura County in southern California, *Saissetia* spends the summer in the adult stage; eggs are laid and hatch in the autumn, and the immature stages are present during winter (September–April); there is only one generation each year (Flanders, 1949). But in the same district, on an oleander which is kept well watered during summer so that it grows vigorously and produces a dense growth near the ground, there may be places among the lower branches where it is moist and cool enough during summer to enable *Saissetia* to continue breeding throughout the hot weather.

The small encyrtid *Metaphycus helvolus* preys on *Saissetia* in its immature stages but is not able to pierce the armor of the mature scale. Now Flanders has observed that the predators, by virtue of a certain pattern of behavior, tend not to penetrate into the dark recesses among the lower branches of these oleanders, so that all stages of *Saissetia* living there are secure from *Metaphycus*. (Compare this natural situation with the model that Flanders made in the laboratory with *Ephestia* and *Habrobracon*, sec. 10.222.) Most of the young *Saissetia* hatching among the lower branches climb upward and settle on the higher branches; they provide a continuous supply of food for *Metaphycus* throughout the year. Very few survive, but, no matter how thoroughly *Metaphycus* seeks them out and eats them, Flanders' observations suggest that the predators may never go seriously short of food because a fresh supply of prey is constantly moving up from below.

In the groves of citrus quite a different sequence of events takes place. No *Saissetia* is hidden away in a place where *Metaphycus* will not go. During winter (September–April) the entire population is in a stage suitable for *Metaphycus*. In the mild climate of the Piru district the predator may have several generations during this period, and it has been known virtually to exterminate *Saissetia* from local areas which may be quite extensive. In these circumstances *Metaphycus* dies out also. In due course the area may be recolonized by *Saissetia*, and in the interval before it is found again by *Metaphycus* the scale insects may become very numerous indeed.

Flanders has suggested that if there are planted adjacent to a grove of citrus, preferably upwind with respect to the prevailing winds, a number of oleanders and if these are kept well watered, then the last event in this sequence may not occur. The populations in the grove of citrus both of *Saissetia* and of *Meta-*

phycus will constantly be replenished from the populations on the oleanders. They will not die out altogether, but neither will they become abundant. Paradoxically, the occurrence in the district of places which are especially suitable for *Saissetia*, both because the humidity permits it to breed continuously and because it is secure from its predator, may result, not as might have been expected, in an increase in the numbers of *Saissetia* but, instead, in a decrease.

12.3   THE DISTRIBUTION AND ABUNDANCE OF PLACES WHERE AN ANIMAL MAY LIVE

12.31   *The Concept of "Carrying Capacity" in the Ecology of Animals Which Recognize Territories*

According to Leopold (1933, p. 136), the requirements of the bobwhite quail *Colinus virginianus* may be summarized under four headings: (*a*) For nesting, it requires well-drained ground covered with moderately thin grass or brush, but with bare ground near by so that the young may dry out after rain. This probably accounts for the frequency with which nests are found near paths and roads. (*b*) The food of the young consists mostly of insects, but as they grow older they also eat seeds and berries; insects make up about 20 per cent of the diet of the adults during summer; seeds are the most important constituent during winter. (*c*) For protection from bad weather and as a place in which to hide from predators, thickets of scrub or tangles of vines are required, especially during winter. (*d*) For sleeping, an elevated place is best, because it enables the birds to take wing readily if they are attacked by a predator. Where meadow, field, scrub, and woods meet, the quail is likely to find an ideal place to live.

Imagine an area of 1 square mile on which these four sorts of country were equally represented but each consisted of solid blocks, occupying one-quarter of the area (see Fig. 12.06, *A*). It is probable that the central point, where all four areas meet, would provide a good place for one covey of quail, but there would be none elsewhere on the square mile. Now imagine the same area of each sort of country arranged patchily as in Figure 12.06, *B*. On this square mile there would be five places where a covey of quail would find all its requirements. Errington would say that the "carrying capacity" of the second arrangement was higher than that of the first. The concept of carrying capacity was developed by Errington (1934) and Errington and Hamerstrom (1936), in the first place, with special reference to bobwhite quail and was later used by Errington (1946) in a more general way to apply to populations of other vertebrates, especially birds and rodents, in which "territorial behavior" was manifest in at least some degree.

Errington and Hamerstrom (1936) found that for the bobwhite quail the carrying capacity of an area was largely determined by the amount and suitability of places for hiding and sheltering during winter, because in most of the places that they studied the presence of farmland insured plenty of food during both winter and summer. During summer there was also plenty of "cover" almost everywhere, with the result that there was room for many more quail during summer than during winter. With the approach of autumn, much of this cover disappeared, and the quails could find adequate shelter only among the

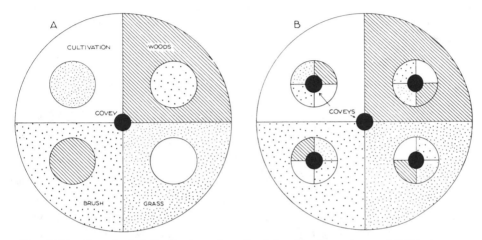

FIG. 12.06.—Hypothetical distributions of woodland, brushland, meadow, and field in two areas, to illustrate the concept of "carrying capacity" for a species like the bobwhite quail, which requires all four sorts of vegetation in the place where it lives. There is the same area of woodland, brushland, meadow, and field in *A* as in *B*; but in *A* their distribution provides a place for only one covey, whereas in *B* there are places for five coveys. (Adapted from Leopold, 1933.)

vegetation growing in gullies, woodlots, or along watercourses. If the natural increase during summer had resulted in numbers in excess of what this amount of vegetation might shelter during the winter, then the surplus was driven out to be destroyed, largely by predators. For 6 years Errington and Hamerstrom counted or estimated the numbers of quails overwintering in 20 different areas in Iowa and Wisconsin. They found that the maximal numbers surviving in a particular area were remarkably constant, and they considered this to indicate the "carrying capacity" of the area. In 4 areas which were studied more intensively, this varied from one quail to 5 acres to one quail to about 65 acres; but the number, whatever it might be, was characteristic of each area (Errington, 1934).

Errington and Hamerstrom (1936) were not able to recognize with precision what were the qualities in the vegetation that made for high or low carrying capacity. There was no doubt that drastic changes in the vegetation providing shelter during the winter changed the carrying capacity substantially; sometimes even minor changes caused marked alterations in the carrying capacity. They wrote (p. 396): "We do know, on the other hand, that carrying capacity has been both raised and lowered by changes in cover conditions. Environmental manipulation, for one purpose or another, may often be of profound consequence to bob-white populations, but we see no way of predicting its effect in advance. To be sure, we may predict with reasonable certainty that a strong spread of fencerow and roadside brush in central Iowa may be followed by an increase of quail and any wholesale reduction of existing cover would probably mean decidedly fewer quail, but there is a great deal about this that we do not know, especially as to details. Leads as good as any perhaps may be given by the data from territories number 11, number 17 and others where relatively slight environmental modifications have made territories unattractive, if not lethal. The cleaning out of one small but strategically located patch of roadside growth apparently changed number 13 from a habitable to a lethal territory. The burning of a single brush-pile, likewise strategically located, patently left territory number 11 with an altered status, so far as wintering quail were concerned." The development of the country for agriculture has, of course, profoundly altered its carrying capacity for quail. Errington and Hamerstrom, taking into account the probable relative distribution of areas providing food and shelter, expressed the opinion that the carrying capacity of this particular region was probably at a maximum about 1880.

In some of the areas studied during some years, the numbers remained below the estimated carrying capacity. Sometimes the number would be below the carrying capacity at the beginning of the winter, despite the great opportunities for multiplication during the summer; sometimes catastrophic shortage of food, severe weather, the presence of other nonpredatory animals, or harrying by men during winter reduced the numbers below the carrying capacity. Also, when the observations were carried on for a longer period (one area was studied continuously for 15 years), the carrying capacity seemed to fluctuate in a way that was not obviously related to changes in the vegetation. These matters will be mentioned again in section 13.34.

The principle of the carrying capacity is also well illustrated in Errington's (1944, 1948) accounts of the ecology of the muskrat *Ondatra zibethicus*. During the winter the muskrat is relatively tolerant of its fellows, and larger numbers can overwinter together than would be tolerated in the same space, once the breeding season has begun. The number left behind in the place where the population spent the winter, after the surplus have been driven forth in the spring, depends largely on how many the muskrats will themselves tolerate in those particular quarters. What proportion of those which are driven out at this time survives to breed depends on their chances of finding adequate quarters before they succumb to predators or one of the numerous other hazards that beset a wandering muskrat. Thus "carrying capacity" has much the same meaning as the number of places in the area where a muskrat may live securely.

The muskrat is semiaquatic; it builds burrows or "lodges" which open below the surface of the water. They are best suited by an expanse of shallow water which does not fluctuate in depth. Marshes are good; so are drains, if they are fed by tiles, especially if they run near cultivated fields which provide good food. Streams also provide places for muskrats to live; but, according to Errington, a stream becomes increasingly suitable, the more it resembles a marsh. Fluctuations in the level of the water may be harmful: floods may drive the muskrats out of their burrows; droughts may uncover the burrows and expose the muskrats to their predators (sec. 12.23). In some situations the level of the water, being dependent on the weather, is likely to fluctuate rather abruptly. This makes it more difficult with the muskrat than it was with the bobwhite quail to think of the "carrying capacity" as a relatively stable quality of a particular area. Nevertheless, Errington (1946, p. 147) has observed that: "The numbers of breeding pairs tolerated in a specific habitat[2] tended to be similar from year to year, irrespective of the less drastic environmental changes, such as moderate fluctuations in water and food."

Temporary changes in carrying capacity may be brought about by the weather, especially fluctuations in rainfall. In Iowa the high level of water sustained in certain marshes as a result of an unusual amount of rain reduced the carrying capacities of these marshes. On the other hand, a series of wet seasons resulted in a big increase in the numbers of muskrats breeding in the state (Iowa) as a whole because the heavy rain filled up ponds, ditches, and marshland which had previously been dry (Errington, 1944).

Changes of a more permanent nature have been brought about by artificially draining, ditching, or damming in areas of marshland and other suitable country. Dymond (1947) referred to the construction of dams, dikes, and canals in the delta of the Saskatchewan River, Manitoba, which diverted water into dried-up marshes and thereby greatly increased the number of suitable places for muskrats to live: the number of muskrats in the area increased from 1,000 to 200,000 in 4 years. Errington (1948) referred to a number of other projects which had permanently increased the carrying capacities of particular areas: "But manipulation of water surely always will be one of the principal techniques in management of muskrat environment, whether the project involves the damming of a ravine for ponds (Allan, 1939), the establishment of a private muskrat 'ranch' (Grange, 1947), or something as ambitious as the restoration of the Lower Souris marshes of North Dakota (Henry, 1939)."

According to Errington (1946, p. 147), food is often less important than "cover" in determining the number of muskrats that an area may support. Nevertheless, when the amount of cover is artificially increased, the point may be reached where food becomes limiting. Errington (1948, p. 598) wrote in connection with this:

2. In this sentence "habitat" seems to mean the actual space occupied by the particular group of muskrats, e.g., a particular marsh in Iowa.

. . . mention should be made that manipulation of water levels on marshlands needs to be directed towards insuring against too much water as well as against too little. The broad-leaved cattail (*Typha latifolia*), a most outstanding native food for marsh-dwelling muskrats of northern United States, seems unable for long to tolerate a depth of more than about 4 feet of water, and it has happened that, when the deep-water stands started dying, those of the shallows likewise died, thus sharply lowering the attractiveness and habitability of given marshes for muskrats. . . . I would estimate that the best cattail marshes allow ascendancies of muskrats to reach nearly twice the populations as do even superior marshes grown to other vegetation. Specifically, the highest wintering density recorded during the Iowa investigations was about 35 per acre for a sizeable area of a cattail marsh, compared with maxima of between 15 and 20 per acre for bulrush (*Scirpus* spp.) marshes. In less extreme cases, one seemingly may expect about as many 20-per-acre populations in cattails as 10-per-acre populations in bulrushes.

The mourning dove *Zenaidura macroura* in North America used to live along the edges of forests, flying to the prairie to feed and to the forest to nest. As the country was developed for agriculture, there came a great increase in the area where forest and farmland met and a corresponding increase in the numbers of mourning doves. But at the present time, in some areas at least, the more intensive use of land for farming and the failure to replace shelter belts and windbreaks is reducing the number of places where *Zenaidura* may live. McClure (1943) recorded the number of *Zenaidura* on three small farms in Iowa during the years 1938, 1939, and 1940. On all the farms and especially on two of them, shade trees were dying and not being replaced during this period. There was a marked reduction in the number of doves from 1938 to 1940, especially on the two farms where more trees died (Table 12.05). McClure

TABLE 12.05*

NUMBER OF NESTINGS AND NUMBER OF YOUNG RAISED BY *Zenaidura macroura* DURING THREE SUCCESSIVE YEARS IN AREAS WHERE NUMBERS OF TREES WERE DIMINISHED

	FARM A			FARM B			FARM C		
	1938	1939	1940	1938	1939	1940	1938	1939	1940
Nestings.............	102	51	37	78	42	31	51	44	46
Successful nestings.....	47	20	15	36	15	10	34	22	21
Young raised.........	80	34	26	62	25	17	61	38	38

* After McClure (1943).

attributed the decrease in the numbers of doves to the reduction in the number of places where nests might be built. The English sparrow, *Passer domesticus*, increased in numbers at all the farms during this period, but the increase was most striking at farm B, where the decrease in doves was also greatest. The sparrow and the dove require much the same sort of food, and a possible (though by no means certain) explanation of this observation would be that the departure of the doves, due to scarcity of places for nesting, left more food available for the sparrows. Considerably more data would be required to verify this hypothesis.

In section 11.11 we mentioned Kluijver's (1951) observations of the great tit *Parus major* in an area of woodland where the places for nesting had been artificially increased so that there were always some boxes left unoccupied; also there was usually plenty of food. Since neither food nor places for nesting were limiting, it may be inferred that the carrying capacity of this area (that is the number of territories which the birds maintained there) depended upon some other more subtle requirement which the birds needed from the place where they chose to live. Although this could not be demonstrated quite so objectively with respect to the bobwhite quail and the muskrat, it is clear from Errington's writings that he had recognized the same phenomenon with these species, too. On the other hand, Kluijver (1951) mentioned certain other areas where the introduction of nesting boxes had been followed by an increase in the numbers of great tits, from which it may be inferred that the carrying capacity of these areas had been chiefly limited by the number of places where a nest might be built. This is quite like the very simple situation we described for the imaginary bees in our hypothetical example in section 2.121.

12.32 *The Concept of "Outbreak Centers" in the Ecology of Locusts and Other Species Which Do Not Recognize Territory*

A few individuals of the locust *Chortoicetes terminifera* may be found during summer (and perhaps at other seasons) if one searches diligently enough, in the right sort of country, scattered throughout most of the continent of Australia (Fig. 12.07). Over most of this enormous region the death-rate is so high that one doubts whether an isolated population would survive there for long. This point cannot be resolved because *Chortoicetes*, like other well-known locusts, migrates freely over great distances in its solitary, as well as its gregarious, phases (sec. 5.4) and areas which may be unsuited for permanent habitation are continuously receiving immigrants from distant places, where the locusts breed more successfully. Occasionally the immigrants come in immense gregarious swarms, and their progeny, if they encounter favorable weather in the area which they have invaded, may breed there for a few generations and constitute a major "locust plague" until such time as catastrophe, in the form of drought and starvation or disease, overtakes them and the plague comes to an abrupt end (Davidson, 1936*a;* Andrewartha, 1937, 1940; Key, 1942, 1943, 1945).

The locusts experience similar hazards in the "outbreak areas," whence these plagues originate, and, during periods of unfavorable weather, r may remain negative, leading to a consequent decrease in the numbers of the locusts in these areas, too. But the outbreak areas are situated in a zone where the climate is kinder to the locusts and catastrophes may be both less severe and less frequent; also opportunities for increase may be greater and more frequent in this climate (see theoretical discussion in chap. 14). Nevertheless, the locusts do not survive equally well or multiply equally readily throughout the whole of the zone experiencing this more favorable climate; they survive and multiply best in certain areas where the particular distribution of certain kinds of soil and vegetation fulfils their rather specialized requirements for egg-laying, feeding, and sheltering. The special qualities of an outbreak area were investigated by Clark (1947*a*), the distribution of outbreak areas in eastern Australia by Key (1945, and earlier papers), and in South Australia by Andrewartha (1940).

In the outbreak area studied by Clark (1947*a*), *Chortoicetes* developed

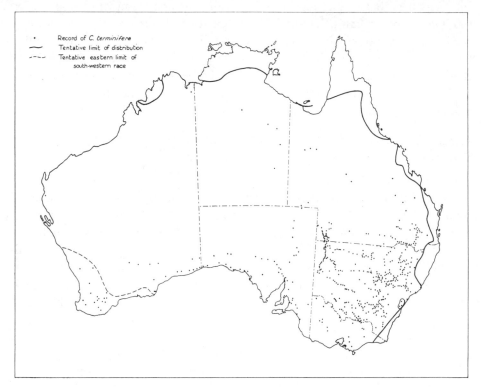

FIG. 12.07.—Localities from which individuals of *Chortoicetes terminifera* have been collected. (After Key, 1954.)

actively only during the warmer parts of the year, from September to May. A few of the older nymphs and adults hibernated successfully, but usually the winter was spent in the egg stage. Between September and May, one, two, or three generations might develop, depending on the amount and distribution of rainfall. Rain was needed to maintain moisture in the soil for the eggs to develop and hatch and to promote the growth of grasses for the nymphs and adults to eat. At least some eggs might remain alive for 3 months during summer in soil that was too dry to promote development, but exposure to drought more prolonged than this would kill most of the eggs. No ordinary drought would kill the hardy, tussocky perennial grasses which dominated local situations in the area, but they would not produce any succulent growth during a severe dry spell. The annuals would die off, leaving only seeds, and the areas which had been dominated by them would remain largely bare while the drought lasted. Much rain would produce a wealth of succulent grass for the locusts to eat, but even after light rain the hardy perennial species would put forth a little succulent growth. Excessive rain (probably not a very frequent hazard in this climate) might engender outbreaks of fungal or bacterial disease among the locusts.

In addition to food, the nymphs and adults required shelter. They would crawl into a tussock of grass to escape from a cold wind at night; during the hottest part of the day they would rest in the shade of the tussock or climb it on the shady side to a position where the breeze might cool them. After unusually good and soaking rains at the right time of the year, there was plenty of shelter provided by a luxuriant blanket of grasses and other herbage not only in the situations normally supporting perennials but also in other areas, where, at other times, the vegetation would at best be low and sparse or even absent. Except after such unusual rains, the only places where adequate shelter obtained were areas which supported perennial grasses with a tussocky habit of growth.

The adults also required, for oviposition, soils of a particular degree of compactness; and it was necessary that these soils should carry, at the time when the eggs hatched, at least sufficient vegetation to provide food and shelter for the early nymphal instars. Soils that were compact enough for oviposition did not support the sort of vegetation which would insure food and shelter in dry weather; and vegetation which adequately provided these requirements was never found on soil which was suitable for oviposition. So it turned out, on investigation, that the distribution and abundance of certain types of soil constituted the most essential qualities of an outbreak area.

The distribution of soils in the area was summarized by Clark (1947a, p. 10): "The Bogan-Macquarie outbreak area has a flat topography. Practically the whole of the present surface is alluvial in origin. The slight local differences in level occurring throughout the area rarely exceed a few feet. However, to them are related major changes in soil type and vegetation. The soils of the higher ground are compact. In general those at the lower level are self-mulching. The latter are regarded as a more recent alluvium than the former. The outbreak area as a whole consists of a mosaic of compact and self-mulching soils. Rarely can more than a few miles be travelled in a straight line in any direction without a major change in soil type being encountered." The heavy self-mulching soils vary in texture from clay loams to clay, as well as in other features; they are all chernozems or rendzinas. The compact soils of the higher levels vary in texture from sandy loam to loam.

All but the most compact of these were favored by *Chortoicetes* for oviposition. The nymphs, hatching from the eggs, might in dry weather find themselves short of food and shelter; but at other times, hatching after rain, they usually found in the low vegetation on these areas sufficient food and shelter to carry them through the first two or three instars. But this vegetation was likely to dry off quickly if no more rain came; moreover, the older nymphs and adults required more adequate shelter than the low sparse vegetation which grew on the compact soils (Fig. 12.08). Their chance of survival in most generations,

FIG. 12.08.—The sort of place favored by *Chortoicetes terminifera* for oviposition. The eggs are mostly laid in the bare ground near the margins of the vegetation. The low sparse vegetation is largely *Chloris truncata* and *Medicago* spp. (After Clark, 1947a.)

and especially during dry weather, depended on their being able to find more reliable sources of food and better places to shelter. The nymphs cannot fly, so they had to find their requirements within a few hundred yards of where they were hatched from eggs.

The characteristic vegetation growing on the heavier soils at the lower levels included perennial grasses, such as *Stipa*, *Eragrostis*, *Chloris*, and *Danthonia*, the tussocks of which provided good shelter for the locusts. These species and others associated with them on these soils of high water-holding capacity remained green and succulent for a relatively long time after rain and thus provided a relatively secure source of food for the locusts (Fig. 12.09).

Fig. 12.09.—The sort of place where the older nymphs and adults of *Chortoicetes terminifera* are likely to find adequate supplies of food and plenty of shelter even during dry times. (After Clark, 1947*a*.)

Wherever these two soils with their concomitant vegetations occur close together, especially if the boundary between them is abrupt, the locust is likely to find its best chance to survive during dry weather and multiply when the weather becomes more clement (Fig. 12.10). This is especially true if the two soils which come into juxtaposition represent the extremes of their classes, because then the vegetation is likely to remain true to type even through extremes of drought. In the area studied by Clark these two soils were distributed as in a mosaic; consequently, there were in this area a great many places where a locust would have a good chance to survive and multiply; the area was extensive—Key (1945) estimated that it contained 10,000 square miles.

These highly suitable places were called "outbreak centers" by Key (1945).

Just as outbreak centers may have a characteristic distribution and abundance in a particular outbreak area, so may outbreak areas be distributed in a particular way in a broader zone whose boundaries are determined by climate. Key plotted the limits of the zone in eastern Australia where the weather would be about as kind to *Chortoicetes* as the weather in the Bogan-Macquarie out-

Fig. 12.10.—A place that is especially suitable for *Chortoicetes terminifera* because of the abrupt transition from the area of compact soil, with its low sparse vegetation which is suitable for oviposition, to the area of heavy self-mulching soil, carrying vegetation which provides a reliable source of food and shelter. (After Key, 1945.)

break area. Then, by studying the distribution of soil and vegetation, he mapped the boundaries of a number of other outbreak areas. Figure 12.11 shows that these are irregularly distributed and that the whole of the area within the uniform climatic zone is far from equally likely to promote the multiplication of *Chortoicetes terminifera*.

Figure 12.11 also includes two outbreak areas in South Australia, which were mapped by Andrewartha (1940). These occur in a broad inland area of semi-desert, where the mean annual rainfall is about 8 inches, compared to 18 inches in the area studied by Clark. The country in the outbreak areas in the desert is characterized by a special topography; the lack of a well-developed drainage system and the presence of local features, such as flat watercourses, flats, and depressions, result in a series of local situations moister than the surrounding countryside, where perennial tussocky grasses persist and where the locust may

FIG. 12.11.—The distribution of outbreak areas for *Chortoicetes terminifera* in eastern and South Australia. (Modified from Key, 1945; and Andrewartha, 1940.)

find food and shelter to enable it to survive during bad times (Fig. 12.12). Elsewhere in the desert, although the climate is little different, the country lacks the essential physiographic features and suitable places, for *Chortoicetes* may be rare or absent. Although the average annual rainfall does not exceed 8 inches, it is highly variable, and the desert occasionally experiences a run of wet years which enables the locusts to multiply. This happens only rarely. The first plague of locusts recorded for South Australia is known to have reached the coast in 1845, less than 10 years after the colony was founded. Since then, plagues have recurred at intervals of about 30 or 40 years. In eastern Australia, plagues have been at least three or four times more frequent than this (Key, 1938). This reflects the difference in climate in the areas where the plagues originate.

Also it is reasonably certain that in eastern Australia the development of the country for sheep-grazing (the chief industry in the area that we have been discussing) has resulted in an enormous increase in the number of places where *Chortoicetes* has a good chance to survive and multiply. It follows that, to a large extent, the locust plagues of this region are man-made; if they have not become more frequent, they must, at least, have become greater since the

Fig. 12.12.—The sort of place where *Chortoicetes terminifera* breed during good years and survive during bad in the outbreak areas situated in the semidesert parts of South Australia. (After Andrewartha, 1940.)

country was settled by Europeans. It will be best to quote the evidence for this directly from Clark (1947a, p. 69):

Several years of intensive field observation and discussions on the development of this part of New South Wales, led the author to conclude that most of the outbreak centres have been produced as a result of human activities, including wholesale ring-barking and clearing of timber, grazing, and the introduction of the rabbit to Australia. In combination with severe droughts, and damage due to trampling by the passage of stock to and from watering places and on stock routes, rabbits have undoubtedly played an important part in facilitating soil erosion, thus rendering considerable sections of the country favourable as oviposition habitats[3] for very many years to come.

Some evidence was gained to suggest that the introduction of sheep to the area led to the establishment of pasture species, some of which are favourable to *Chortoicetes*, e.g., *Hordeum leporinum*. The area was first used for cattle grazing, and several of the older property holders, whose statements may be considered reliable, informed the author that when the country was first settled the pasture was much sparser on both the light and heavy soil types than it is today. It was not until after the introduction of sheep that relatively dense pastures

3. "Habitat" in this passage refers to the special sort of place which serves not for the whole life of the animal but for one of its particular functions, e.g., oviposition, feeding, sheltering.

developed, mainly as a result of extensive tree destruction. Thus not only have oviposition nuclei been created, but the food-shelter habitat has been improved and extended as a result of human occupation of the country.

In western Australia and South Australia the development of certain tracts of country for agriculture has similarly resulted in a great increase in the area suitable for the grasshopper *Austroicetes cruciata*. The ecology of this species is discussed in section 13.12.

CHAPTER 13

Empirical Examples of the Numbers of Animals in Natural Populations

Fluctuations occur in every group of animals and in every habitat that has been in-
vestigated: an impression not derived solely from deliberate research upon fluctuating
populations, since it has also been widely supported by ecologists studying popula-
tions for other purposes and from other points of view. These fluctuations are rather
forcibly expressed in human affairs through the irregularity from year to year of
pests, epidemics, and of various animal resources, notably in the sea. . . . . Al-
though the amplitude of fluctuation is often very great, scarcity alternating with high
abundance every so many years, two things that we might expect do not often happen.
The first, complete destruction of vegetation by herbivorous animals, has already been
mentioned. The second is complete destruction over any wide area of either predators
or prey. The factors controlling the limitation of these fluctuations are therefore of
great interest, since they are the factors that critically affect the survival or extinction
of species, ...

ELTON (1938, pp. 130–31)

13.0 INTRODUCTION

ALL the examples which are discussed in this chapter have several qualities in common: (*a*) each one represents an attempt, by one means or another, to measure the number of animals constituting a natural population; (*b*) most of them have been continued through several generations, some up to 15 years or more; (*c*) in each case the investigator has had a more or less thorough knowledge of the behavior and physiology of the species to guide him in his selection of facts about the particular natural population which he has chosen to study; (*d*) in each case the investigator has also studied the environment and related the observed numbers (or more often changes in the numbers) to measurements of the environment. In other words, these are all particular ex- amples of the general method which we extolled in section 1.2—that is, all except the examples given in section 13.4. There is an extensive literature about the so-called "regular cycles" in the numbers of fishes, game birds, rodents, and other fur-bearing animals; and nearly all of it suffers from the grave dis- advantage that information about numbers has been gathered without much reference to the biology of the animals themselves; this was mostly because the knowledge did not exist. Most of the examples in this chapter are species

145

which, for some reason or another, are of importance to man. They have been selected not for this reason but because they have been studied intensively. In some examples man has radically altered their ecology; but whether this has happened or not, the principles which emerge are the same.

Whereas it must surely be generally true that an animal's chance to survive and multiply depends on all four of the major components of its environment (sec. 2.2), yet it is also true that there are plenty of cases, perhaps the majority, in which one or several of the components are of chief importance; if their influence is understood, the others may, in practice, be safely disregarded.

13.01   *Methods of Sampling*

If the chief purpose of an investigation is to explain the influence of environment on the animal's chance to survive and multiply, it may be sufficient to measure relative changes in the density of the population without troubling to ascertain the absolute numbers in the population that is being studied.   Simple trapping may suffice for this purpose (sec. 13.113). In the absence of a method of trapping which can be relied upon to catch a sufficiently constant proportion of the population at all times, it may be necessary to make an estimate of the absolute numbers. Errington (1945), in his study of the bobwhite quail (sec. 13.34), found it practicable to enumerate every bird in the area. But with most species it is not practicable to count the individuals on any area large enough to be of use in a practical study, and some method of sampling becomes necessary. With relatively sedentary species, like the arthropods in soil, which were studied by Salt and Hollick (1946; see sec. 13.02), the numbers may be counted precisely in an adequate number of small samples chosen from a larger area. With more mobile species and those which are not so easily seen or counted even in a small area, the special method known as capture, marking, release, and recapture is most useful if it is practicable (sec. 13.33). On the other hand, even mobile species, if they are easily seen, may be counted on a number of small areas or "quadrats"; this method was used successfully by Clark (1947a) to estimate the relative numbers of locusts on different quadrats.

In the case of the wireworms mentioned in section 13.02, it was sufficient, having chosen a suitably large and uniform area of grassland, to take a number of samples at random over the whole meadow. But suppose one wanted to estimate the numbers of a certain caterpillar in a certain area of mixed woodland. If this caterpillar were known to live only on oak trees, then it would be a rather futile and wasteful procedure to sample all the trees in the wood at random. It would be better to count the number of oaks in the area and then take a random sample of oaks on which to count the caterpillars. This means, in more general terms, that it is usually necessary to place a restriction on the randomness of the sampling within the whole area chosen for study, by taking into account the patchy distribution of suitable places for the animals to live in.

13.02 Tests for Randomness in the Distributions of Animals in Natural Populations

We mentioned in section 2.12 and elsewhere that the places where each sort of animal may live are usually distributed unevenly and that this imparts a fundamental patchiness to the distributions of animals. Furthermore, the animals are usually distributed quite unevenly through the places that are suitable; some places will be crowded, while similar ones contain unexpectedly few individuals. The prevailing scarcity of most species results in many suitable places being quite empty, purely by chance. Other causes of uneven distributions have also been mentioned. For example, with living plants the ordinary processes of growth, senescence, and decay and of succession may result in constantly changing distributions of food and suitable places to live (sec. 5.0). Also the presence of an active predator tends to accentuate the patchiness of the distribution of its prey (secs. 5.3 and 10.321). We give later several examples in which the degree of unevenness in the distribution of the animals in a natural population has been estimated quantitatively.

The best way to find out whether the individuals of a particular species living in an area are distributed over it more or less evenly than they would be if their distribution were random is to partition the area into a number of stations or "quadrats" and count or estimate the numbers occurring in each quadrat. It is necessary to choose a large number of small quadrats, so that the chance that a particular individual will occur in a particular quadrat is small. Then if the animals are distributed at random over the whole area, the frequencies of the quadrats containing 0, 1, 2, 3, . . . , x individuals should tend toward a Poisson series. The frequencies in a Poisson series are given by the expression

$$Ne^{-m}\left(1,\ m,\ \frac{m^2}{1\times 2},\ \frac{m^3}{1\times 2\times 3},\ \cdots,\ \frac{m^x}{1\times 2\times 3\times \ldots x}\right),$$

where N is the total number of animals counted, m is the mean number per quadrat, and the series 0, 1, 2, 3, . . . , x, represents the number of animals that may be found in any one quadrat. It follows that the frequency of quadrats containing x individuals is given by

$$Ne^{-m}\left(\frac{m^x}{x(x-1)(x-2)\ldots 1}\right).$$

In a Poisson series the mean equals the variance, so that

$$\frac{\Sigma(x-\bar{x})^2}{\bar{x}(n-1)} = 1.$$

If, in an observed series, $\Sigma(x - \bar{x})^2/\bar{x}(n - 1)$ is significantly less than unity, this indicates that the variance of the distribution must be less than that appropriate to a Poisson series, which means that the individuals in the population are distributed over the area more evenly than would have been expected from a mere random scattering. Conversely, if this quantity is found to be significantly greater than unity, then it may be concluded that the variance in the population that was sampled was greater than that which is appropriate to a Poisson series; in other words, the animals were distributed less evenly (that is, with greater patchiness) than would have been expected from a mere random scattering. If the number of samples is large (several hundred or more), the theoretical Poisson series may be calculated, and the observed distribution compared directly with the theoretical one. If, as more often happens, the number of samples is smaller, it is appropriate to make use of the fact that, for a Poisson series, $\Sigma(x - \bar{x})^2/\bar{x}$ is distributed as χ^2, with the number of degrees of freedom one less than the number of samples. Alternatively, for larger samples it is convenient to make use of the fact that the variance of the expression $\Sigma(x - \bar{x})^2/\bar{x}(n - 1)$ (for a Poisson series) is $2n/(n - 1)^2$ and therefore observed values of $\Sigma(x - \bar{x})^2/\bar{x}(n - 1)$ may be considered to differ significantly from unity, at the 5 per cent level of probability, if the difference exceeds $2\sqrt{2n/(n - 1)^2}$.

Salt and Hollick (1946) divided a seemingly uniform square yard of grassland into 81 "quadrats," each 4 inches square, and counted the wireworms (larvae of *Agriotes* sp.) in each quadrat. The soil was dug up, taken away, and washed through a series of sieves by a special method which separated nearly all the larvae, even the smallest, from the soil. The results are shown in Tables 13.01 and 13.02. The value of $\Sigma(x - \bar{x})^2/\bar{x}(n - 1)$ calculated from

TABLE 13.01*

NUMBERS OF WIREWORMS IN 81 QUADRATS IN SQUARE YARD OF SOIL UNDER
GRASS

0	3	8	9	4	2	1	2	3
7	3	7	6	3	1	3	1	3
3	1	8	1	4	3	4	3	6
0	1	4	2	11	5	8	1	3
5	1	6	11	9	7	3	9	6
3	2	7	13	2	7	8	9	14
2	0	7	8	15	8	12	7	5
8	3	8	5	6	12	5	3	2
7	8	10	10	11	15	8	10	10

* Each square represents a quadrat in the relative position in which it occurred in nature.
After Salt and Hollick (1946).

these results was 6.24. With 81 samples, a value of this magnitude indicates a highly significant departure from the theoretical Poisson series: the wireworms were definitely not distributed at random in the square yard; there were many quadrats with many or few wireworms and not enough with moderate numbers. In other words, the wireworms showed a strong tendency to occur in patches.

Salt and Hollick also studied the distribution of wireworms in one area of $\frac{1}{4}$ acre of grassland and another of 8 acres. From the former, 21 samples were taken from each of 16 "stations" over a period of 3 years. The samples consisted of cylinders of soil, 4 inches in diameter and 12 inches deep. The "stations" were separated from one another by 5 yards in one direction and 15 in the other. In the 8-acre area, 20 stations were established, and these were sampled at monthly intervals for 28 months. In both areas the wireworms were distributed less evenly than would be expected on the hypothesis of random distribution, and the discrepancies were highly significant.

TABLE 13.02
DATA FROM TABLE 13.01 REARRANGED TO SHOW FREQUENCIES WITH WHICH
DIFFERENT NUMBERS OCCURRED

No. of Wireworms	No. of Quadrats	No. of Wireworms	No. of Quadrats
0.	3	9.	4
1.	8	10.	4
2.	7	11.	3
3.	14	12.	2
4.	4	13.	1
5.	5	14.	1
6.	5	15.	2
7.	8	16+.	0
8.	10		
Total.		461	81
Mean.		5.69	
$\dfrac{\Sigma(x - \bar{x})^2}{\bar{x}(n - 1)}$		2.48	
$2\sqrt{\dfrac{2n}{(n - 1)^2}}$		0.318	

In the $\frac{1}{4}$-acre area there were relatively fewer wireworms where the loam was deeper and fewer in the wetter situations. In the 8-acre meadow there were fewer where the soil contained a lot of organic matter, and these were usually also places where *Lolium* was growing thickly. This indicates that at least some of the patchiness of the numbers of wireworms in these larger areas was associated with variability in the soil with respect to its suitability as a place for wireworms to live. The small area of 1 square yard might be expected to be less variable than the larger areas measured in acres. So one might expect that most of the patchiness in the numbers of wireworms in an area as small as this might be attributed to the behavior of the insects.

Cole (1946*a,b*) placed a large number of boards of the same size on the ground in several sorts of woodland. He examined the boards at intervals and recorded the numbers of the different species of animals that were living under each one. In one experiment he collected 127 centipedes (*Lithobius forficatus*) from 1,152 boards; they were distributed among the boards as in Table 13.03. For these data $\chi^2 = 1,258.6$; $n = 1,151$. With such large values of

TABLE 13.03*

DISTRIBUTION OF 127 CENTIPEDES (*Lithobius forficatus*)
UNDER 1,152 BOARDS PLACED ON GROUND IN WOOD

No. of Centipedes per Board	No. of Boards	
	Observed	Poisson Series
0	1,039	1,031.8
1	101	113.7
2	10	6.3
3	2	0.2

* After Cole (1946*a*).

n, one makes use of the fact that $(\sqrt{2\chi^2} - \sqrt{2n - 1})$ is distributed normally with unit standard deviation about a mean of zero. In this case $(\sqrt{2\chi^2} - \sqrt{2n - 1}) = 2.09$. So the deviations from the theoretical distribution, though small, must be judged significant at the 5 per cent level of probability. There were too many boards with 0, 2, or 3 centipedes and too few with 1; this indicates a patchiness in excess of what would be expected on the hypothesis of random distribution of the centipedes throughout the wood. Similarly, with the isopod *Trachelipus rathkei* and the beetles of the family Carabidae, Cole found that their distributions were significantly nonrandom and patchy. In fact, the only animals in these studies of which the distribution was not significantly nonrandom were certain spiders. Cole also analyzed data from Beall (1940) for the caterpillar *Loxostege sticticalis* and for fleas on rats and found in each case that the distributions differed significantly from the Poisson series and that the difference was due to a patchiness in excess of what would be expected to occur by chance.

Gilmour *et al.* (1946) measured the distribution of the blowfly *Lucilia cuprina* in an area of grassland near Canberra. It was a circle 8 miles in diameter. The country was undulating; most of the original vegetation of savannah woodland had been destroyed, and the area had been developed for grazing sheep; a plantation of pines encroached into the area in one sector. One hundred and two traps were spaced evenly, ¾ mile apart, over this area, as indicated in Figure 13.01. The traps were examined daily, early in the morning before the flies became active, and the numbers of flies in the traps were recorded. The same experiment was repeated three times during the summer of 1941–42, on

December 7, January 18, and March 13. The traps were in the same situations on each occasion.

The authors of this paper very kindly lent us their original records, and we have analyzed them, first, to see whether the individuals of *L. cuprina* occurring naturally on this area were distributed at random over it and, second, to see whether the concentrations of flies recurred in the same local situations in each experiment.

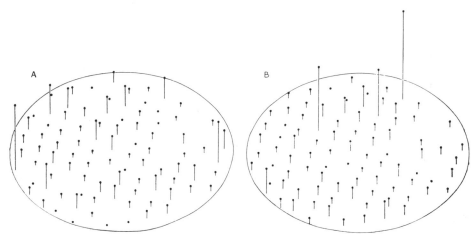

FIG. 13.01.—The distribution of the blowfly *Lucilia cuprina* in an area of grassland near Canberra. The diameter of the circle was 8 miles. Traps were distributed evenly over the area, as indicated by the columns. The height of each column is proportional to the number of flies caught in the trap. Two experiments are represented: *A* was started on December 7; *B*, on March 13. (Data from Gilmour *et al.*, 1946.)

Gilmour *et al.* (1946) had already shown that the number of flies caught in the traps represented between 0.001 and 0.0001 of the total population on the area, from which we may conclude that the chance that any one fly would be caught in a trap was quite small. Therefore, if the flies were distributed at random over the area, the numbers caught in the traps might be expected to conform to the Poisson series. We calculated the theoretical Poisson series for each experiment and compared the theoretical frequencies with the observed ones. The results are set out in Table 13.04. If the flies had been distributed over the area truly at random, values of χ^2 as large as those that may be calculated from the bottom row of the table would be expected less often than once in 1,000 trials. We can therefore say with a high degree of assurance that *Lucilia cuprina* were not distributed at random over this area of approximately 50 square miles. The values of $\Sigma(x + \bar{x})^2/\bar{x}(n + 1)$ show that the departures from randomness were due to excessive patchiness. The same may be inferred by inspection of the frequencies set out in the body of the table; for example, in the second experiment, too many traps caught from 0 to 12 flies or from 23 to 113 flies, and too few traps caught the intermediate numbers of flies. Similar discrepancies between the observed and theoretical frequencies occurred in experiments 3 and 4.

The statistics shown in Table 13.04 establish beyond reasonable doubt that the flies were nonrandomly distributed over the experimental area at the time

TABLE 13.04

OBSERVED FREQUENCIES (TRAPS) COMPARED WITH APPROPRIATE POISSON SERIES FOR EACH OF THREE EXPERIMENTS WITH *Lucilia cuprina* BY GILMOUR *et al.* (1946)

EXPERIMENT No. 2			EXPERIMENT No. 3			EXPERIMENT No. 4		
No. of Flies per Trap	No. of Traps		No. of Flies per Trap	No. of Traps		No. of Flies per Trap	No. of Traps	
	Obs.	Expected		Obs.	Expected		Obs.	Expected
0– 12	52	11.64	0– 3	55	18.26	0– 4	54	19.80
13– 14	7	13.48	4	10	15.12	5	9	13.44
15– 16	6	18.13	5	8	17.17	6	6	15.13
17– 18	1	19.04	6	3	16.25	7	8	14.59
19– 20	3	16.06	7	3	13.17	8– 9	5	21.55
21– 22	4	11.11	8–64	23	21.66	10–88	16	14.29
23–113	28	11.55	. . . .	. . . .	. . . .	. . . .	. . . .	. . . .
Total	207	207	. . . .	99	99	. . . .	81	81
P.	0.001			0.001			0.001	
$\dfrac{\Sigma(x - \bar{x})^2}{\bar{x}(n-1)}$ . .	22			13			22	

of each experiment. An inspection of Figure 13.01 suggests that the distribution of flies had changed markedly from one experiment to the next. To test this hypothesis, we correlated the numbers of flies caught in each trap during the different experiments. The values obtained were $r_{23} = 0.18$, $r_{24} = 0.08$, $r_{34} = 0.11$. None of these values was significant; hence it may be concluded that there was no consistency in the way that the flies were distributed over the experimental area at the times of the different experiments. In other words, the concentrations of flies occurred in different places at different times.

An important feature of these experiments was that 40,000 flies which had been reared in an insectary and marked with a dye were liberated near the center of the circle one day before the trapping began for each experiment. Some of the marked flies were subsequently recaptured in the traps. In each experiment there was a tendency for the traps that caught most wild flies to catch most marked ones also, and vice versa. From this it may be inferred that the local situations where the wild flies were most numerous were also the places that were most attractive to the marked ones. From the correlations given in the preceding paragraph, we know that these centers of concentration were differently situated in December, January, and March. They were therefore unlikely to be associated with permanent physical features like streams, dams, sheds, shade belts, and so on. Nor was there any evidence that they were associated with flocks of sheep.

Dobzhansky and Pavan (1950) measured the relative density of various species of *Drosophila* in a forest in Brazil. Baits were set at 10-meter intervals along two transects, each 200 meters long, which intersected in the middle at right angles. The baits attracted flies in the immediate vicinity, and, at regular intervals during the day, flies that came to the baits were collected by sweeping a net over the baits. The numbers collected reflected the density of flies in the vicinity of the baits. The results for two groups of species are illustrated in Figure 13.02. The lines at right angles represent the transects, and the height

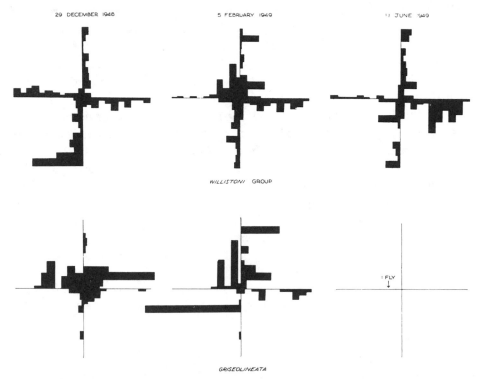

FIG. 13.02.—The distributions of two groups of sibling species of *Drosophila* in a forest in Brazil. Two rows of traps were arranged in the form of a cross, and the experiment was repeated on dates indicated. For further explanation see text. (After Dobzhansky and Pavan, 1950.)

of the black columns along these lines are proportional to the percentage of flies collected at each point. Tall columns indicate that a large proportion of the total flies along the transect occurred at this point; low columns signify that the species were rare. These and other results showed that the distributions of the flies in the forest were extremely patchy. The pattern of patchiness varied from month to month; in places where one species was numerous, the others were often scarce; in other words, the pattern varied for each species. These points are illustrated in Figure 13.02. The causes of patchiness among these populations may have been quite complex. Dobzhansky and Pavan

(1950) suggested that it was associated with irregular distributions of fruits of different sorts which attracted the different species of flies. The flies are also sensitive to moisture; this varied as the transect crossed a damp depression or a dry slope. The flies are also sensitive to light; this varied, depending largely on the amount of vegetation. But whatever may be the chief causes of patchiness, this is a good example of patchiness within a relatively small area. We have no doubt that the phenomenon would have been more marked if the transects had extended for miles instead of meters.

Many sorts of insects and other animals concentrate in small areas for courting or mating. Barton-Browne (personal communication) has observed adults of the fruit fly *Dacus tryoni* congregating at dusk toward the end of an orchard nearest the setting sun. During the brief period of dusk (about 30 minutes) the flies court and mate. They do this at no other time of the day, when they are more widely distributed throughout the orchard.

So far the examples all refer to relatively small areas chosen because they possessed a certain uniformity; the floor of a forest, the soil in a meadow, even the 60 square miles of grassland studied by Gilmour *et al.*, may all be regarded as small uniform areas, arbitrarily chosen from the much larger area which is the distribution of the whole species. Let us now consider a much broader sort of area. Take, for example, the distribution of certain butterflies in Great Britain. Ford (1945*a*, p. 268), drawing from his wide experience of many species, wrote: "Colonies of a species may be localised by a variety of conditions. They may be cut off on islands, or they may be restricted to a particular type of habitat, such as mountains, marshes, moors or woods, and any of these may be completely isolated from places of a similar character. It is probable that in Great Britain the course of civilisation is increasing the isolation of colonies, . . ."

The numbers in each local colony fluctuate, perhaps between wide extremes, and not always synchronously. This adds another component to the variability in the relative abundance of a species from place to place when broad areas are concerned. A quotation from Ford (1945*a*, p. 268) illustrates this point:

". . . fluctuations in numbers and in areas of distribution take place throughout the whole population of certain butterflies, but isolated colonies of a species are especially susceptible to such fluctuations, whether as a part of the more general trend or in response to local conditions. In an isolated habitat the food supply may partly fail, circumstances may for a season or more favour the parasites or the diseases of an insect, or bad weather may occur at some critical stage in the life-cycle. Any of these events may greatly reduce an isolated population and cause it to fall far below the numbers which the area could support in favourable conditions. When these return, a recrudescence of the colony may be expected, so that the colony will fluctuate over a long or short period according to circumstances.

Records of the relative abundance of the butterfly *Euphydryas aurinia* in one isolated colony in Cumberland are available for 55 years. These were summarized by Ford (1945a, p. 268) as follows: "The species was quite common in 1881, and gradually increased until by 1894 it had become exceedingly abundant. After 1897, the numbers began to decline, and from 1906 to 1912 it was quite scarce. From 1913 to 1919 it was very rare, so that a few specimens only could be caught each year as a result of long-continued search, where once they were to be seen in thousands. From 1920 to 1926, a very rapid increase took place, so that by 1925 the butterfly had become excessively common, and so it remained until we ceased our observations in 1935."

The colony in this place, though greatly depleted at times, survived. Elsewhere colonies have been known to become extinct. And this is a risk that is quite important for small colonies that fluctuate widely in numbers. Another quotation from Ford (1945a, p. 135) illustrates how this may add to the patchiness of the distribution of the species at any one time and adds to the variability from time to time: "This butterfly [*Euphydryas aurinia*] occurs widely in suitable places in England, though it largely avoids the eastern counties. However, it used to occur near Deal in Kent, and it has recently reappeared in Hertfordshire after a long absence. Formerly it existed in a few places in East Anglia, where it seems to be extinct, and there are one or two localities for it in Yorkshire."

The best example that we can give of a population distributed with excessive uniformity comes from the work of Holme (1950) with the small lamellibranch *Tellina tenuis*, which lives in vertical burrows in the sand on beaches in Britain. Holme studied the distribution of *T. tenuis* in certain selected areas of beach in the estuary of the River Exe. He chose a number of "stations" situated between high- and low-water marks and counted the numbers of *T. tenuis* in from 2 to 12 quadrats of 0.01 square meter from each station. Altogether, there were 33 samples, which is scarcely enough for an adequate test. The values of $\Sigma(x - \bar{x})^2/\bar{x}(n - 1)$ at three stations, where 4, 12, and 7 samples were taken, were, respectively, 0.13, 0.31, and 0.09. With so few samples, none of these figures is independently significant. But all the 33 samples may be pooled to give an estimate of χ^2. This was found to be 6.437, which, with 22 degrees of freedom, corresponds to a probability of less than 0.01. The observed distribution is significantly different from the Poisson series, but in this case the variance is too small; such a degree of uniformity in spacing would have happened by chance less often than once in 100 trials.

It is not known whether *Tellina tenuis* has some means of "defending a territory" or whether this uniformity in its distribution comes about in some other way. In an area as uniform as a few square yards of beach, "territorial behavior" on the part of the animals living there might be expected to lead to a high degree of uniformity in their distribution. With birds living in an area where resources of food, nesting sites, shelter, and so on may be unevenly distributed, territorial behavior may not lead to any marked uniformity in the distribution of the birds.

Allee *et al.* (1949, p. 366) and Hutchinson (1953) mentioned several other examples in which the pattern of distribution of certain natural populations had been studied. It is most unusual to find one in which the distribution is random. It is almost equally unusual to find one in which the departure from randomness is in the direction of excessive uniformity. It is generally true, with very few exceptions, that natural populations are distributed nonrandomly and that the departure from randomness is in the direction of excessive patchiness. This has to be taken into account when devising methods for sampling natural populations and for testing the significance of differences between means (see Bliss and Fisher, 1953). Also, of course, it emphasizes the unreality of those mathematical models which assume that populations of animals are distributed either at random or uniformly through any considerable area of country.

13.1 OF FOUR NATURAL POPULATIONS IN WHICH THE NUMBERS ARE DETERMINED LARGELY BY WEATHER

The numbers of many animals are largely determined by weather. In chapters 6, 7, and 8 we have shown how the main components of weather, which are temperature, moisture, and light, may influence the animal's chance to survive and multiply. In this section we describe how weather determines the numbers in certain natural populations. Studies of this sort must be based on a sound knowledge of the physiology and behavior of the animal, but it is not necessary to add much of these details to this chapter, since this has already been done in earlier chapters.

In the first example in this section, weather is shown to influence the animal's chance to survive and multiply directly and also indirectly through its influence on the amount and distribution of food. In the second example, weather is important, in addition to its direct influence, indirectly through its association with food and diseases. In the last two examples the association between weather and the numbers in the populations is clear enough, but the relationships have not been explained so fully as in the first two examples. Weather may also come into the discussions of the examples in sections 13.2 and 13.3, but it does not have such a dominant place as in those selected for discussion in this section.

13.11 *Fluctuations in the Numbers of* Thrips imaginis *in a Garden in South Australia*

A natural population of *Thrips imaginis* in the grounds of the Waite Institute near Adelaide, South Australia, was studied by Davidson and Andrewartha (1948*a*, *b*). This insect belongs to the order Thysanoptera, family Thripidae. It is indigenous to southern Australia, and, like many other species in this region, its opportunities for multiplication have increased greatly since the country was developed for agriculture. But this era of change is largely over now.

13.111 BIOLOGY OF *Thrips imaginis*

The life-cycle of *Thrips imaginis* is not complicated by diapause, nor is there any season of the year when it becomes quiescent. Breeding and development go on continuously, but it will be shown later that both the birth-rate and the survival-rate fall to low levels during the height of the summer and during the winter, so that r remains negative during these seasons of the year.

The fully grown insect is about 1 mm. long, and many hundreds of them can find room in one rose. In nature they frequent the flowers of roses, fruit trees, a wide variety of garden plants, and weeds. In the laboratory, adults lived for 77 days at 24° C. on a diet that lacked pollen; but they laid scarcely any eggs; 9 thrips averaged 20 eggs each. When pollen was added to the diet, the average length of life was decreased to 55 days, but the average number of eggs increased to 209. On a diet which included pollen the mean length of life varied from 250 days at 8° to 46 days at 23° C. The mean number of eggs per thrips varied from 192 at 12.5° to 252 at 23° C. Eggs were produced fairly evenly throughout adult life at the rate of 1.4 per day at 12.5° and 5.6 per day at 23° C. (Andrewartha, 1935).

In nature the eggs are usually imbedded in the soft tissues of the flower, and the two nymphal stages are commonly found in the flowers alongside the adults. In the laboratory it proved impossible to rear the nymphs on a diet that lacked pollen; but when pollen was added to the diet, they could be reared quite readily. The duration of the life-cycle from egg to adult was 44 days at 11° and 9 days at 25° C. (Andrewartha, 1936). A generation may be completed in a few weeks during the warm period of the year but may require many months during the winter (Davidson, 1936*b*).

Not only is the speed of development greatly retarded during the winter, but the death-rate, especially among the nymphs, may be high because few flowers would last long enough for the nymphs to complete their development at this season of the year. The stages best adapted to survive the winter are the adults and the pupae. The latter are sometimes found in the flowers alongside the active stages, but it is more usual for the fully grown nymph to leave the flower and pupate among the litter at the base of the plant or in the soil just below the surface. The egg, imbedded in the tissue of the plant, is surrounded by moisture; the nymphs and adults may replenish their supplies of moisture by sucking sap; but the pupae, which do not feed, are likely to be fatally desiccated during hot, dry weather. This may be one of the major causes of death during the summer.

The behavior of the adult with respect to dispersal was not observed directly. Nevertheless, there was clear evidence that the species has a strong innate tendency toward dispersal; this evidence was summarized in section 5.1. This behavior is advantageous during the spring, when there are suitable flowers everywhere. But a strong tendency to disperse when flowers are very sparsely distributed may result in a high death-rate among the adults during the height of the summer and during the winter (sec. 11.12).

During this investigation samples of flowers containing *Thrips imaginis* were examined almost daily for 14 years. We occasionally found a few other species of animals in the flowers. These included several other species of thrips, an occasional beetle, a few stray aphids or mites. All told, they would add up to a fraction of 1 per cent of the numbers of *T. imaginis* in the flowers. A few adult *T. imaginis* might occasionally be caught in spiders' webs, though certainly this was an infrequent occurrence. A few pupae might have been eaten by foraging beetles, spiders, and mites, etc.; but we gained no evidence of this, and, if it occurred at all, it must have been on quite a small scale. Indeed, *T. imaginis* is remarkable for the virtual absence of other sorts of animals from its environment.

13.112   CLIMATE AND VEGETATION IN THE AREA WHERE THE POPULATION WAS LIVING

The climate of Adelaide is broadly like that of the region near the Mediterranean Sea. The winter is mild; the mean minimal and maximal temperatures for July (coldest month) are 7.5° and 15° C., respectively. The summer is hot and rather severely dry; the mean minimal and maximal temperatures for January (hottest month) are 16.3° and 30° C., respectively. The little rain that falls during summer is quite inadequate to moisten the parched earth except very temporarily (Fig. 6.04). The more permanent vegetation consists mainly of xerophytic shrubs and trees, which remain dormant during summer; the annuals, which constitute nearly all the herbaceous vegetation of the area, spend the summer as seeds. Most of the rain falls during the winter; usually there is sufficient to maintain the soil permanently moist for about 7 months of the year (Davidson, 1936c; Fig. 11.01). This is the growing season for plants, and it culminates in a great outburst of blossoming in the spring.

The ending of the summer and the beginning of the growing season is quite abrupt, and one can usually recognize the "break" of the season quite unequivocably. During the 15 years between 1932 and 1946 the earliest "break" came on February 18 (1946) and the latest on June 3 (1934). When the break came early, the seeds germinated quickly, and the plants completed relatively more of their vegetative development during the autumn before growth was slowed down by the lower temperatures of winter. In these circumstances the plants flowered earlier in the spring and more abundantly. For example, one annual *Echium plantagineum*, which is commonplace and of which the flowers are highly favored by *Thrips imaginis*, was observed in full bloom on August 3 in 1946 but not until October 3 in 1936; other species responded similarly. It will be shown later (sec. 13.115) that this has a big influence on the numbers of *T. imaginis* present during the spring.

13.113 METHOD OF SAMPLING

Simple trapping was practiced, and only adults were recorded. Simple trapping may indicate relative changes in abundance but provides no way of estimating the absolute numbers in the area drawn upon by the traps. In this case the traps consisted of 20 roses picked at random from a long hedge in the garden; the roses were always chosen at the same stage of development. The particular variety, "Cecil Brunner," lacks anthers, so they are useless as breeding places for *Thrips imaginis*. But the thrips are strongly attracted to them, and they are excellent traps for the adults. From April, 1932, to December, 1946, the roses were collected at 9:00 A.M. every day except Sundays and certain holidays; from 1939 to 1946 daily samples were collected each year during the period September to December, except that no records are available for 1944.

Simple trapping gives an adequate estimate of relative abundance only if the traps catch a sufficiently constant proportion of the population independent of variations in numbers and other circumstances. In the present experiments there was clear evidence of rather pronounced short-term fluctuations superimposed on a steady trend either up or down, depending on the season (Fig. 13.03). This suggested that the numbers which we were catching on any one day depended partly on the numbers in the area and partly on the activity of the thrips in seeking out the flowers. We have no elaborate information about the behavior of the *Thrips imaginis* to guide us in seeking an explanation of this varying activity. But experience suggested that the thrips were likely to be more active on a fine warm day. Such weather often occurs in Adelaide when the barometer is falling.

The hypothesis was tested by fitting the data to a partial regression of the general form

$$\log \frac{Y}{Y_t} = b_1(x_1 - X_1) + b_2(x_2 - X_2) + b_3(x_3 - X_3) \dots,$$

in which Y and Y_t refer to the numbers of thrips; x_1, X_1, x_2, X_2, . . . , refer to certain components of the weather; and b_1, b_2, b_3, . . . , are coefficients which measure the strength of the association of the dependent variate log Y/Y_t with the independent variate $(x_1 - X_1)$, $(x_2 - X_2)$, . . . , independent of their association with each other. The numbers of thrips were converted to logarithms because it was more instructive and realistic to consider relative changes in their numbers than absolute changes. The dependent variate was expressed as the logarithm of a ratio, and the independent variates as differences, because it was necessary with all of them to eliminate trends with time before testing the hypothesis about activity. In the case of the thrips, the trend was due chiefly to the growth of the population; with the independent

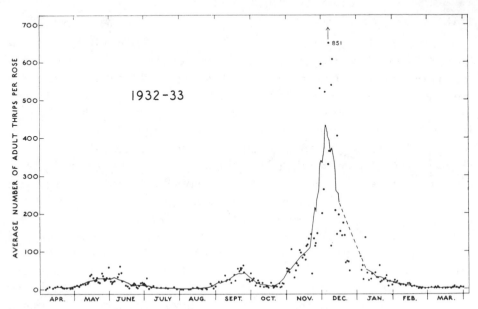

FIG. 13.03.—The individual daily records and the trend throughout 1 year in the number of *Thrips imaginis* per rose. The points represent daily records. The curve is a 15-point moving average. (After Davidson and Andrewartha, 1948a.)

variates, the trends were due to the progression of the seasons. The trends were eliminated by fitting the data to equations of the general form

$$Y = a + bx + cx^2 + dx^3.$$

Hence in the equation for partial regression given above, Y_t, X_1, X_2, X_3, . . . , refer to values taken from the calculated smooth trend lines and Y, x_1, x_2, x_3, . . . , refer to the actual daily records of thrips, temperature, rainfall, and so on (Fig. 13.04).

In the equation for partial regression log Y/Y_t may also be written (log Y − log Y_t), which serves to emphasize the fact that in this method we are comparing departures from the usual, or expected, number of thrips with departures from the usual, or expected, weather. Thus if there are unexpectedly many thrips in the roses on days that are unusually warm, this will be indicated by the value of the appropriate coefficient. Similarly, if there are unusually few thrips found on unusually wet days, this will be measured too.

Moreover, if wet days are also sometimes cold, this method will tell to what extent the low numbers recorded on wet days are associated with rain independently of temperature and to what extent with low temperature independently of rain.

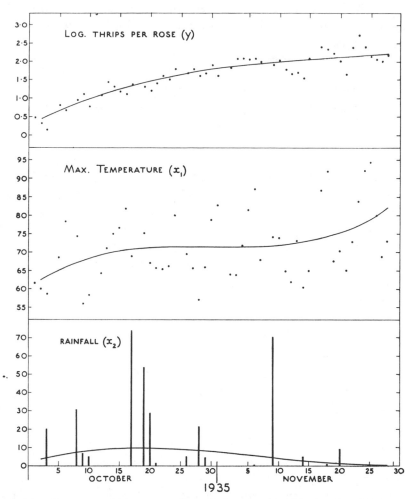

Fig. 13.04.—The curves which were used to eliminate trends with time in relating activity of thrips, log y, to maximum temperature, x_1, and rainfall, x_2. For further explanation see text. The curves were calculated from the expressions:

$$Y = 1.6195 + 0.04288\,\xi_1 - 0.005235\,\xi_2 + 0.00006393\,\xi_3;$$
$$X_1 = 71.805 + 0.3359\,\xi_1 + 0.008789\,\xi_2 + 0.003627\,\xi_3;$$
$$X_2 = 5.715 - 0.2056\,\xi_1 + 0.06132\,\xi_2 + 0.002255\,\xi_3;$$

in which X_1 is in degrees Fahrenheit; X_2 in units of 0.01 inches of rain; and Y in units of log thrips per rose. (After Davidson and Andrewartha, 1948b.)

The first partial regression which was formulated included three independent variates—for rainfall, temperature, and barometric pressure. The last was shown to have no influence independently of temperature and rainfall. So it was excluded from the regression. Since the roses may be attractive to *Thrips imaginis* for several days before they are picked, the next regressions included six independent variates, namely, temperature and rainfall for each of the three days preceding the collection of the roses. The regression as a whole was significant, but only those terms relating to the day immediately before sampling were significant in their own right. So a third regression was calculated with only two terms. It was

$$\log Y - \log Y_t = 0.007925(x_1 - X_1) - 0.001969(x_2 - X_2),$$

where x_1 was the maximal temperature in degrees Fahrenheit and x_2 was rainfall in units of 0.01 inch for the day immediately before that on which the sample was taken.

From this equation it may be shown that, on the average, the samples overestimated the numbers of thrips by 2.5 per cent for every degree the temperature exceeded the "usual" as indicated by the trend line. Similarly, the samples underestimated the numbers by 6.6 per cent for every 0.1 inch of rain. The regression accounted for only 10 per cent of the total variance, so there was probably much "activity" of the thrips which we had failed to explain. Nevertheless, this equation may be used to correct the crude results given by the samples. All the figures used in the calculations discussed in section 13.115 have been so corrected. Details of the method may be seen in Davidson and Andrewartha (1948*b*).

13.114   THE DATA

Samples were taken on 2,291 days throughout the 14 years, and, of these, 1,252 samples were taken during the period from September 1 to December 30, which is the season when the thrips multiply to a maximum each year. Altogether, during the whole investigation some 6,000,000 thrips were recorded from the samples.

These data were analyzed to discover how closely certain components in the environment of a *Thrips imaginis* were associated with the maximal numbers attained each year. The method of partial regression was used. The dependent variate was a quantity which represented the number of thrips; the independent variates were quantities which measured certain components of the weather. The first step was to calculate from the data a quantity which would represent the abundance of the thrips during the period of each year when they reached their greatest numbers. Figure 13.05 shows that the numbers of thrips in the roses were always low at the beginning of the spring; they increased during the next few months and then declined more or less abruptly. Although the daily fluctuations were pronounced, the trend, as indicated by a 15-day running average, was clear, and the maximum of the curve could be clearly recognized. The date on which the maximum occurred was usually close to November 30, but it varied from November 15 in 1939 to December 13 in 1945. The dates on which the maxima occurred each year are indicated by arrows in Figure 13.05. The maximum of the smoothed curve might have served as the criterion we were seeking; but, for reasons which were explained in Davidson and Andrewartha (1948*b*), we chose instead the average for the 30 days preceding the maximum. The mean logarithm of the numbers of thrips in one rose for these 30 days for each of the 14 years are shown in Table 13.05.

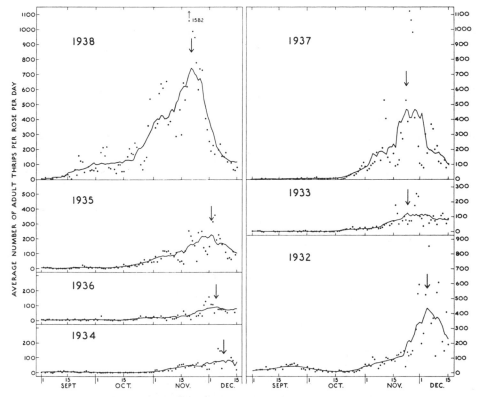

FIG. 13.05.—The numbers of *Thrips imaginis* per rose during the spring each year for 7 consecutive years. The points represent daily records; the curve is a 15-point moving average. The arrow indicates the date on which the curve reaches the maximum for each year. Note that the same general trend was repeated during the spring each year, with variations in the height attained by the curve and date on which the maximum was attained. (After Davidson and Andrewartha, 1948*a*.)

These are the values which constituted the dependent variate (log *y*) in the regression, which is discussed in section 13.115.

In our choice of independent variates for the regression we were guided by our knowledge of the biology of *Thrips imaginis* and the climate of the area where the population was living. In this climate, *r* for *T. imaginis* is likely to be negative at all seasons of the year except during spring and perhaps for a brief period during autumn in some years (Figs. 13.03, 13.05). The explanation for this may be given briefly as follows: During the summer, except for the early part, the places where pupae occur are likely to be so arid that few may be fortunate enough to survive; also, suitable flowers for breeding become scarce and sparsely scattered, and this adds to the hazards of life for adults, reducing their chances both of surviving and of leaving offspring behind them. During the winter, food continues to be sparsely distributed; also with the prevailing low temperature the thrips develop slowly, and many nymphs die simply because they fail to complete their development before the flowers in which they are living wither. With rising temperatures in the spring, the thrips develop more quickly; the soil, wet from winter and remoistened by spring rains, permits a high rate of survival among pupae. In some years flowers continue to be scarce at first, but it is a striking feature of this region that the blossoming of a high proportion of the natural vegetation is condensed into a brief period in the spring; the annuals, which preponderate, must set seeds before the summer drought sets in; and the perennials also tend to produce most of their flowers at this season. Consequently, the thrips flourish and multiply at this time until, as summer develops, the soil dries out and the flowers wither and disappear. The thrips which had been so abundant become few again.

Even when the thrips are most numerous, the flowers in which they are breeding do not appear to be overcrowded except perhaps locally or temporarily: while the thrips are multiplying, the flowers increase even more rapidly. Then, when the population begins to decline, the flowers become less crowded still. This may be partly because the survival-rate among pupae depends upon the moisture in the top half-inch of soil, whereas the plants draw their water from a greater depth. Also, as the flowers begin to thin out and the distances between suitable breeding places become greater, r may be reduced by this cause long before the flowers become absolutely scarce (sec. 11.12).

Considerations of this sort led to the hypothesis that the numbers achieved by the thrips during each year were determined largely by the duration of the period that was favorable for their multiplication. When this period was prolonged, the thrips would ultimately reach higher numbers; when it was briefer, the decline would set in while the numbers were still relatively low.

It was explained in section 13.112 that the flowers of a wide variety of annual plants are of chief importance for *Thrips imaginis*. The date on which these plants burst into flower depends largely on the stage in vegetative growth reached by the end of the winter. The date on which blossoming finishes depends largely on the amount of rain falling during the spring. The beginning of blossoming is more variable than its ending, but both may influence the duration of the favorable season for *T. imaginis*.

Therefore, in seeking quantities which might be associated with the numbers of thrips in the spring, we looked first for one which would represent the opportunity for growth during autumn and winter afforded to the annual plants which were chiefly important in the ecology of *Thrips imaginis*. This was done by the familiar method of summing "effective temperature" in units of day-degrees. Assuming arbitrarily a threshold of 48° F., the number of day-degrees of effective temperature appropriate to each day were calculated from the expression

$$T = \frac{\text{Maximal daily temperature} - 48}{2}.$$

By reference to certain climatological studies of this region (Davidson, 1936c; Trumble, 1937) we devised an arbitrary method of determining the "break" in the season; this was taken as the date on which the seeds began to germinate. Starting from this date and proceeding to August 31, daily effective temperatures were summed. The 14 yearly totals are given under the heading x_1 in Table 13.05.

TABLE 13.05*
INDIVIDUAL VALUES OF QUANTITIES POSTULATED AS LIKELY TO INFLUENCE NUMBERS OF THRIPS
PRESENT DURING SPRING

YEAR	DATE OF "BREAK" OF SEASON	INDEPENDENT VARIATES†				DEPENDENT VARIATE‡
		x_1	x_2	x_3	x_4	LOG y
1932	Mar. 27	12.09	4.37	6.09	13.77	2.08
1933	Apr. 11	10.42	4.39	7.04	12.09	1.57
1934	June 3	6.33	6.55	7.07	10.42	1.65
1935	Mar. 14	13.92	5.48	7.15	6.33	2.03
1936	Apr. 7	11.54	3.94	6.48	13.92	1.54
1937	Mar. 28	12.37	4.37	7.16	11.54	2.04
1938	Feb. 19	18.74	2.03	7.29	12.37	2.51
1939	Feb. 24	18.34	2.71	6.82	18.74	2.72
1940	Apr. 7	10.89	2.38	7.85	18.34	2.10
1941	Apr. 3	12.96	6.48	6.81	10.89	2.40
1942	Mar. 30	12.62	6.05	7.28	12.96	2.57
1943	Apr. 7	9.63	4.36	6.96	12.62	1.79
1944	Apr. 2					
1945	May 4	8.33	6.76	6.42	9.65	1.93
1945	Feb. 18	17.69	3.28	6.43	8.33	2.28

* The numbers of thrips are given as log y in this table. After Davidson and Andrewartha (1948b).
† x_1 = Total effective day-degrees (in units of 100 day-degrees) from the "break" of the season to August 31. x_2 = Total rainfall (in inches) for September and October. x_3 = Total effective day-degrees (in units of 100 day-degrees) for September and October. x_4 = As for x_1 but for the year preceding the one when the samples of thrips were taken.
‡ y = The geometric mean number of thrips in one rose for the 30 days preceding the maximum of the smoothed curve (see explanation in text).

Rain falling during the spring would tend to sustain the plants and thus prolong the period favorable for the multiplication of thrips. Also rain falling at this time might be expected to increase the rate of survival of pupae. So the total rainfall for September and October was included in the hypothesis; it appears as x_2 in Table 13.05.

Although, as a rule, the spring is warm enough to promote the breeding of *Thrips imaginis*, the temperature may be lower than optimal. Therefore, we included the temperatures during the spring in the hypothesis and calculated as x_3 the sum of the daily effective temperatures for September and October.

Finally, there was the possibility that the influence of a "good" season might be carried forward to the next by some means or another, perhaps through a greater "carry-over" of thrips or of seeds. To test this possibility, x_4 was added to the quantities in Table 13.05. It is the same quantity as x_1, but the series has been displaced by 1 year.

13.115 ANALYSIS OF THE DATA

The method of partial regression described by Fisher (1948, p. 147) is well suited to the analysis of data of this sort, because each coefficient of partial regression expresses the degree of association between its independent variate and the dependent variate when all other terms *included in the regression* are held at their means. In other words, it measures the degree of association which each one has with the dependent variate independently of all the others. Regression measures degree of association; causal relationships need to be inferred on biological grounds. This is generally true of all correlations, but the point may be driven home in relation to the present study by supposing that some other quantity *not included in the regression* were correlated closely with, say, x_2. Then if this hypothetical quantity were substituted for x_2 in the calculations, it would be found to have the same degree of association with y as x_2 had. But we would still be left to decide which one (if either) was causally related to y; and this would have to be done from our knowledge of the biology of the animal and its environment. The biology of *Thrips imaginis* had been investigated thoroughly in preparation for this ecological study; and the climate of Adelaide was well known from the work in climatology that had been done at the Waite Institute.

Before giving the results of the analyses, it is necessary to explain why the numbers of thrips were transformed to logarithms. We would expect that for a certain increment in, say, temperature the numbers of thrips would be multiplied by a constant factor; we would not expect their numbers to be increased by the addition of a constant number. This may be expressed in symbols: the expected relationship would not be of the form $Y = k + ax$ but would more likely be of the form $Y = ka^x$; or, if several "causes" were being investigated, it would be

$$Y = ka_1^{x_1} \cdot a_2^{x_2}. \tag{i}$$

Taking logarithms of this expression gives

$$\log Y = \log k + x_1 \log a_1 + x_2 \log a_2. \tag{ii}$$

This equation is, indeed, the general equation for partial regression, which is usually written in statistical textbooks in the form

$$Y = a + b_1 x_1 + b_2 x_2. \tag{iii}$$

Therefore, it was appropriate merely to transform the numbers of thrips to logarithms and proceed with the analysis by the standard methods of partial regression. This was done, giving the equation

$$\log Y = -2.390 + 0.1254x_1 + 0.2019x_2 + 0.1866x_3 + 0.0850x_4. \qquad \text{(iv)}$$

From equation (iv) the expected (or theoretical) values for y may be calculated. These are shown in Figure 13.06. The agreement with the observed values is

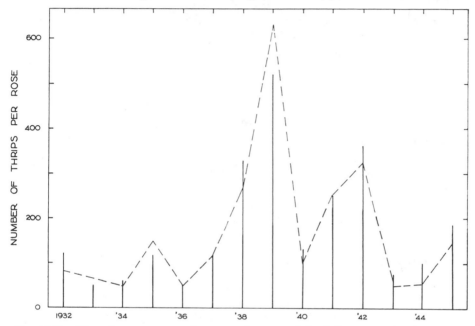

Fig. 13.06.—The columns represent the geometric means of the daily counts of thrips per rose. The curve represents the theoretical values for the same quantities calculated from the expression:
$$\log Y = -2.390 + 0.125x_1 + 0.202x_2 + 0.187x_3 + 0.058x_4.$$
The independent variates x_1, x_2, x_3, and x_4 are the quantities given in Table 13.05. Note that they are altogether different quantities from those designated x_1 and x_2 in Fig. 13.04. (After Davidson and Andrewartha, 1948b.)

remarkable. But it is more instructive for our present purposes to present the results in tables setting out the analysis of variance (Table 13.06b) and the significance which can be attributed to the regression and the individual terms in it (Table 13.06a).

The relative magnitudes of the coefficients $\log a_1$, $\log a_2$, and likewise the percentage increases associated with unit increases in x_1, x_2, . . . , depend upon the arbitrary choice of units for x_1, x_2, . . . Therefore, the magnitudes of these quantities provide no guide to the relative importance of x_1, x_2, . . . , in determining the magnitude of $\log y$. This difficulty may be overcome by expressing the coefficients in "standard measure" thus:

$$\beta_1 = b_1 \sqrt{\frac{S(x_1 - \bar{x}_1)^2}{S(y' - \bar{y}')^2}},$$

TABLE 13.06a
REGRESSION OF LOG y (NUMBER OF THRIPS PER ROSE) ON x_1, x_2, x_3, AND x_4 (SEE TABLE 13.05)

| VARIATE | REGRESSION COEFFICIENTS | | S.E. | t | P |
	β	b			
x_1............	1.224	0.1254	0.0185	6.79	< 0.001
x_2............	0.848	.2019	.0514	3.93	$< .01$
x_3............	0.226	.1866	.1122	1.66	$< .2 > 0.1$
x_4............	0.511	0.0580	0.0185	3.08	< 0.02

Increase per unit increase in x_1 = 41.7 per cent,
Increase per unit increase in x_2 = 67.1 per cent,
Increase per unit increase in x_3 = 62.0 per cent,
Increase per unit increase in x_4 = 18.9 per cent.

where b_1 stands for log a_1 and y' stands for log y. The values of β_1, β_2, . . . , are given in the first column of Table 13.06a. They show that x_1 was easily the most important of the independent variates and that x_2 was the next most important. The untransformed regression coefficients, b_1, b_2, . . . , are also shown in Table 13.06a, because they are used in the tests of significance. The analysis of variance is set out in Table 13.06b.

TABLE 13.06b
ANALYSIS OF VARIANCE

Source of Variance	Degrees of Freedom	Sum of Squares	Mean Square	Variance (Per Cent)
Regression..........	4	1.5537	0.3884	78.4
Residual............	9	0.2736	0.0304	....
Total...........	13	1.8273	....	....

$z = 1.2739$; $P < 0.001$.

It was not practicable to make a similar precise analysis of the numbers of thrips which were recorded during the periods when they were at a minimum. At times the thrips averaged less than one per flower; and parallel samples taken from other sorts of flowers in which the thrips were breeding indicated correspondingly low numbers. The sampling variance of such low numbers is large, making quantitative analysis difficult.

But there are, nevertheless, several general conclusions which may be stated with reasonable certainty. Since, during times of low numbers, the thrips were always very few in every flower which was examined, we can say categorically that they were not short of food in the sense that all the available food was consumed; on the contrary, only a small fraction of the food was used. On the other hand, food certainly was scarce (or perhaps better to say "sparse") in the sense we discussed in section 11.12. It is obvious that shortage of food (in this sense) cannot be interpreted as a "density-dependent factor."[1] We can

1. The phrase "density-dependent factors" recurs in this chapter. We use it with the precise meaning in which it is defined in sec. 2.12 in a quotation from Elton (1949).

also be reasonably certain, as we stated in section 13.111, that predators and diseases are virtually absent from the environment of *Thrips imaginis*. Nor was there any other observation made which could be taken to indicate the operation of a "density-dependent factor."

On the other hand, by analogy with what is known to happen during the season of maximal abundance, we can hazard a well-informed guess as to what might be determining the numbers during the season of minimal abundance. During the summer the rate of increase becomes and remains negative, largely because of the lack of moisture. Aridity causes a high rate of deaths among pupae and a scarcity of places for breeding. There is a brief respite in the autumn, after the rains come, and r sometimes becomes positive again for a brief period. During the winter, r becomes negative once more, and the chief cause of this is low temperature. The numbers might therefore be expected to decline further, the longer and more arid the summer and the longer and colder the winter.

13.116   CONCLUSIONS TO BE DRAWN FROM THE ANALYSIS OF THE DATA

During 14 years the numbers of this natural population of *Thrips imaginis* fluctuated with a consistent seasonal rhythm (Figs. 13.03 and 13.05). The maximal numbers were usually reached about the end of November; there was a minimum at the end of the summer about March–April, followed by a small increase, which reached its peak about May or June; then, finally, another decrease, during which the lowest numbers of the year would be registered, usually toward the end of July or during August.

The maximal numbers varied from year to year, as illustrated in Figure 13.06, but at no time were the thrips numerous enough to consume more than a small fraction of the food or occupy more than a small fraction of the total places available to them. So their numbers were never limited by shortage of food in the absolute sense. Nevertheless, the degree of concentration of breeding places was important because the thrips do not multiply rapidly except when they are closely surrounded by suitable flowers for breeding. Indeed, it was shown that a substantial part of the variation in the maximal numbers from year to year could be explained by reference to the duration of the period when breeding places were distributed densely over the area. This was largely determined by weather. Other causes for the fluctuations in the maximal numbers of thrips from year to year were also related to the weather; and, altogether, 78 per cent of the variance was explained by four quantities which were calculated entirely from meteorological records.

This left virtually no chance of finding any other systematic cause for variation, because 22 per cent is a rather small residuum to be left as due to random sampling errors. All the variation in maximal numbers from year to year may therefore be attributed to causes that are not related to density:

not only did we fail to find a "density-dependent factor," but we also showed that there was no room for one.

The evidence with respect to what determined the numbers during the season of minimal abundance is not quite so definite. It was shown beyond doubt that food never acted as a "density-dependent factor" and with reasonable certainty that no other "density-dependent factor" was operating. It was suggested, by analogy with what is known of the influence of weather on maximal numbers, that weather also determined minimal numbers in a way that was essentially independent of density. The maximal and minimal numbers for the seasonal cycles were studied in these ways, but the average for the whole year was not calculated, because this would have little meaning and would not be relevant to the discussion of "density-dependent factors."

Those who believe that the numbers in natural populations can be "regulated" only by "density-dependent factors" argue that, in the absence of "density-dependent factors," populations must either go on increasing without limit or else become extinct. And we have, indeed, been asked by our colleagues, in discussing these data, why *Thrips imaginis* does not become extinct and why *T. imaginis* does not go on increasing without limit. The answer to the first question is that in many local situations the thrips do die out every year. But in the broader area which we were studying, there were always some less severe places where there was never time enough for the population to become extinct. In them, r was negative also, and the population was declining there, as elsewhere, but not rapidly enough to die out before the weather changed and permitted the population to start increasing again.

To make this point doubly clear, consider what happens near the margins of the distribution of *Thrips imaginis*. As one travels north from Adelaide, one eventually approaches the central desert. But long before the desert is reached, there comes a region where *T. imaginis* may not be found, although there is little doubt that if a population of *T. imaginis* were to be set down in a favorable place in this region during spring, it would survive for a while. But it would not survive for long, probably not even through its first year, for in this region the summer is more arid and more prolonged than in Adelaide; less relief is provided by the autumn; and droughts may also occur during the winter. In this area, therefore, a population of *T. imaginis*, after the spring had passed, would be likely to decrease so rapidly, and the decline would be likely to continue for so long, that the population would have little chance of surviving long enough to take advantage of the arrival of spring. Consequently, *T. imaginis* is not found there. But in the area that we studied there were always many local situations which were sufficiently favorable even in summer for *T. imaginis* to survive there.

The answer to the second question as to why *Thrips imaginis* does not go on multiplying without limit can be given with equal simplicity. It just does not have time to do this, for the favorable season of the year, which is the spring, is invariably followed by an unfavorable period in the summer, when hot, dry weather (but not "density-dependent factors") knocks the numbers back.

The dogma of "density-dependent factors" is unrealistic on at least two major counts: it ignores the fluctuations of r with time, which may be induced by seasonal and other fluctuations in the components of environment; it also ignores the heterogeneity of the places where animals may live. This empirical study of a natural population of *Thrips imaginis* has shown that if these two facts are recognized, it is not necessary to invoke "density-dependent factors" to explain either the maximal or the minimal numbers occurring in a natural population.

13.12   *The Distribution and Abundance of* Austroicetes cruciata *in South Australia*

The ecology of the grasshopper *Austroicetes cruciata* was studied by a team of workers at the Waite Institute during the period 1935–42. The work was summarized by Andrewartha (1944*b*). Particular aspects of the work were reported by Andrewartha (1939, 1943*a*, *b*, 1944*a*, *c*), Andrewartha, Davidson, and Swan (1938), Birch (1942), Birch and Andrewartha (1941, 1942, 1944), and Andrewartha and Birch (1948).

13.121   THE LIFE-CYCLE OF *Austroicetes cruciata*

The life-cycle of *Austroicetes cruciata* is characterized by an intense obligate diapause in the egg stage (secs. 4.3 and 4.6). Consequently, there is one and only one generation each year; also the same stage in the life-cycle recurs at the same season each year, with remarkably little variation in the dates on which they begin and end. Diapause is usually completed by about midwinter (June). The speed of development of the eggs thereafter depends largely on the temperature experienced during the latter part of the winter and early spring (sec. 6.234). Variability in the temperature at this season induces variability in the date on which the eggs hatch; but, for one district, it was estimated, from the records of temperatures for the years 1891–1940, that the peak of the hatching varied by no more than ± 10 days about the average date for 80 per cent of the years. The other 20 per cent of occasions included departures of up to 20 days on either side of the average date. During the few years for which direct observations are available, the date varied from August 25 to September 16.

This is the season of the year when the herbage attains its greatest development (sec. 4.6), and, except on the occasion of an unusually severe drought, the nymphs usually find plenty of food. The speed with which they develop is therefore determined almost entirely by temperatures. The nymphal development is condensed into a rather brief period, and the adults lay their first batch of eggs within a few days of the final molt. According to our observations, the nymphal stage occupied from 41 to 54 days, and the date on which the first eggs were laid varied from October 30 to November 3.

By November, summer is already advancing, and the grass is beginning to wither. As their food disappears, the adults of *Austroicetes cruciata* begin to die from starvation long before most of them would have died from old age. The date by which they had virtually disappeared varied from November 20 to December 3. From then until the following spring (September) the whole population remained in the egg stage. During the whole of the summer the eggs remained firmly in diapause, which confers upon the eggs a remarkable capacity to withstand the severe drought characteristic of the summer in this area (sec. 7.233).

The regularity of the life-cycle with respect to the seasons and the fact that virtually only one stage was present at a time made it easier to ascertain the probable influence of weather on the grasshopper's chance to survive and multiply (sec. 13.124). The behavior of the females in seeking places for oviposition was described in section 12.22; the distribution of soil that is suitable for oviposition has an important influence on the distribution and abundance of *Austroicetes cruciata* (Andrewartha, 1944b).

13.122   CLIMATE AND VEGETATION IN THE AREA WHERE *Austroicetes cruciata* LIVES

The grasshopper *Austroicetes cruciata* lives in an area where the climate is more arid than that in the region where *Thrips imaginis* is found in abundance (sec. 13.112). For example, the mean annual rainfall in the district of Hammond, which is in the heart of the distribution of *A. cruciata*, is 11.7 inches; about 62 per cent of this falls during the cooler half of the year (May to October). The growing season in this district lasts, on the average, for about 5 months. During the warmer half of the year (November to April) 4.6 inches of rain falls, on the average. But during the same period the evaporation from a free-water surface amounts to about 31 inches. The soil remains parched and the plants dormant (Figs. 13.07 and 11.01).

The rainfall in this region is variable with respect to both the amount that falls each year and its distribution throughout the year. So the average values do not indicate the full severity of the hazards experienced by *Austroicetes cruciata*. From the records for 33 stations in the area where *A. cruciata* lives, it was calculated by Cornish (in Andrewartha, 1943a) that, on the average, once in 20 years the total annual rainfall would be nearly double the mean and once, on the average, during the same 20 years it would be little greater than half the mean. In other words, the rainfall for the wettest year in any group of 20 years is likely to be 3.4 times as great as the rainfall for the driest year of the same 20-year period; the extremes in any 10-year period are likely to differ from each other in the ratio 2.5:1, and for any 5-year period in the ratio 1.8:1. These figures give only an indirect measure of the variability of the rainfall; but since both excessive dryness and excessive wetness may, in certain circumstances, result in the deaths of many *A. cruciata*, it is clear that variability of this magnitude must add to the hazards of life for the grasshopper.

The temperature during the winter in the area where *Austroicetes cruciata* lives is low enough, especially at night, to promote diapause-development but not to harm the eggs. But the summer may be extremely hot. The mean maximal temperature in the shade for January (the hottest month) at Hammond is 32° C. The mean maximal temperature of the soil at about ¾ inch below the surface, where the eggs of *A. cruciata* spend the summer, is likely to be at least 10° C. higher than this (sec. 6.234); and, on unusually hot days, the maximal temperature of the soil at this depth must exceed 50° C. But there is evidence, both from experiments in the laboratory and from observation in nature, that very few eggs ever die from the direct influence of high temperatures, provided that they do not at the same time lose a fatal amount of water.

The pristine vegetation of this country was woodland and scrub, with a sparse understory of xerophytic shrubs and grasses. There must have been relatively little area suitable for the multiplication of *Austroicetes cruciata*. Most of the original vegetation has now been destroyed, and its place has been taken by pastures composed of a variety of hardy grasses and other low-growing herbs which serve as food for grasshoppers. Most of them are annuals, which are adapted to take advantage of the brief growing season characterizing this region: they become established from seed during winter while the soil is moist; they put forth a great burst of growth during spring, and complete the production of seeds in the brief period that elapses before the summer drought sets in. Barley grass, *Hordeum murinum*, is typical and is one of the most important foods of the nymphs. The adults depend more on the grasses which stay green a little longer. The growth of several perennial species of *Stipa*, *Danthonia*, and others follows the same seasonal pattern as the annuals, but they shoot afresh each year from a dormant crown; also they will put forward a small showing of green leaves at any time after a few points of rain. These and some of the longer-lived annuals are important as food for the adults. Also in isolated local situations small areas sown to wheat or lucerne may continue to support a local population after grasshoppers have mostly disappeared elsewhere.

13.123   THE NUMBERS OF *Austroicetes cruciata*

The numbers of *Austroicetes cruciata* fluctuated between wide extremes. From 1935 to 1939 the swarms were too obvious to need counting; the individual swarms did not seem to increase in density, nor did the boundaries of the area affected by swarms expand, but there was a perceptible increase from 1935 to 1939 in the proportion of this area occupied by the swarms. By the end of October, 1940, the grasshoppers had become so scarce that hardly any could be found, although we searched most thoroughly. Virtually this entire generation died without laying an egg, and in 1941 the numbers were still extremely low. A few sparse populations were found, but the great majority of the places which were examined seemed to be quite devoid of *A. cruciata*.

The distribution of *Austroicetes cruciata* was determined precisely, but the numbers only qualitatively. We traveled along roads and tracks running north and south and east and west, stopping every 5 or 10 miles in places which seemed suitable for the grasshoppers and recorded their numbers in five categories which we called "absent," "solitary," "dense solitary," "loose swarm," and "swarm." Apart from local fluctuations such as the disappearance of swarms from certain districts after the hot, dry summer of 1938–39, there was no appreciable change in the boundaries of the area where the grasshoppers were numerous during the period 1935–39. We also went back as far as 1891, searching the technical publications and newspapers for references to plagues of *A. cruciata*. There was no reference to swarms outside the boundaries of the area indicated by our surveys, although there was a number of references to swarms inside these boundaries. The results are shown in Figure 13.07.

We also studied the causes of deaths in the different stages of the life-cycle. We were successful in this, partly because the life-cycle of *Austroicetes cruciata* is so regular with respect to the seasons, but also because we fortunately encountered at one time or another during the course of the investigations extremes of weather during summer, winter, and spring.

13.124   THE CAUSES OF DEATHS AMONG NATURAL POPULATIONS OF *Austroicetes cruciata*

The egg, being present during summer, autumn, and winter, is exposed to a number of different sorts of hazards from weather at different seasons; but the only one that matters is the risk of losing a fatal amount of water during summer. The fully turgid diapausing egg at the beginning of summer contains about 2.4 mg. of water. It can lose more than half this and still remain alive; but if its water content falls below about 0.75 mg., the egg is likely to die. The membranes which inclose the tissues of the egg are remarkably impermeable to water passing outward, and eggs that were stored for seven months at 20° C. and 55 per cent R.H. lost water to the air at the rate of about 0.01 mg. per day. In another experiment, four batches of eggs were stored for 210 days at 30° or 35° C. at 55 per cent and 75 per cent. R.H. The death-rate varied from 32 to 59 per cent. Although the passage of water outward through the egg membranes to dry air is so slow, it can nevertheless be taken in rapidly from damp soil. Eggs which had been desiccated until they were flaccid were placed on moist paper at 25° C. During the first 20 hours they absorbed water at the rate of 0.21 mg. per hour, for the second 20 hours at the rate of 0.10 mg. per hour, and after 60 hours they had become fully turgid again.

In nature during the summer the diapausing egg is losing water most of the time but regains some after each substantial shower of rain. If the rain is sufficient to keep the top inch of soil moist for several days, the egg will have time to replace all the water that it has lost. During the summer of 1938–39 the drought was unusually severe, and in certain districts a high proportion of

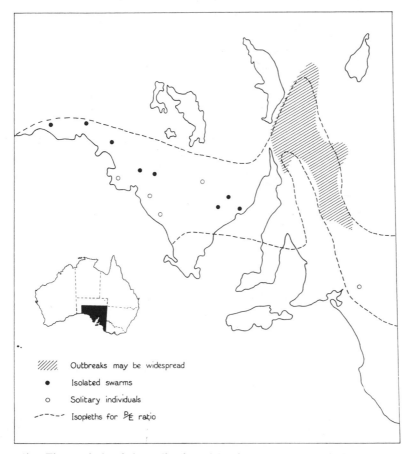

Fig. 13.07.—The area in South Australia where *Austroicetes cruciata* may, during a run of favorable years, maintain a dense population. For further explanation see text. (After Andrewartha, 1944*b*.)

eggs lost a fatal amount of water. We measured the proportion of eggs that died in the two districts of Hawker and Hammond, which are separated by about 40 miles. At Hawker the proportion of dead eggs varied from 94 to 88 per cent depending on the local situation, and at Hammond from 75 to 54 per cent (sec. 12.22). We measured the intensity of drought associated with these death-rates by estimating from records of rainfall, temperature, and atmospheric humidity the amount of evaporation that would have occurred from a free surface of water during the longest spell without rain. At Hammond there were 80 consecutive days without rain, at Hawker 89; the estimated evaporation during these periods was 26.2 inches at Hammond and 40.6 inches at Hawker. Assuming that the severity of drought at Hammond during the summer of 1938–39 was just about critical, we estimated the number of occasions that this had happened during the 50 years between 1891 and 1940. We found that at Hammond there had been 11 such occasions, giving a probability of 22 per cent for any one generation of a high death-rate from drought in the summer. If a more severe criterion were chosen, say, an estimated evaporation of 40 inches, then this probability would be about 10 per cent (Table 13.07). The figures in Table 13.07 show that the risk from drought during summer is

TABLE 13.07*

PARTICULAR YEARS BETWEEN 1891 AND 1940 FOR WHICH DROUGHTS AS SEVERE AS THOSE AT HAMMOND AND HAWKER DURING 1938–39 WERE RECORDED FOR THREE DISTRICTS IN AREA WHERE *Austroicetes cruciata* LIVES

YEAR	HAWKER		HAMMOND		ORROROO	
	Duration of Longest Dry Spell (Days)	Total Evaporation (Inches)	Duration of Longest Dry Spell (Days)	Total Evaporation (Inches)	Duration of Longest Dry Spell (Days)	Total Evaporation (Inches)
1892/93	107	37.2	...	...	109	34.9
1897/98	133	45.0	...	...	113	44.1
1898/99	...	...	91	39.1	...	...
1900/01	...	...	81	32.7	119	37.6
1905/06	110	39.6	117	40.1	105	36.0
1914/15	...	...	73	30.7	...	...
1915/16	116	39.6	141	51.4	96	28.4
1918/19	102	42.9	...	...	106	41.9
1921/22	114	37.9	99	41.3	76	26.5
1922/23	...	...	88	40.6	69	29.9
1924/25	...	...	78	33.3	...	...
1925/26	96	40.7	...	...	...	...
1928/29	154	71.1	154	50.5	85	32.3
1932/33	74	28.8	124	49.4	...	...
1938/39	89	40.6	80	26.2	...	...
1939/40	66	34.1	...	...	...	...
No. years evaporation > 26.2 in.	11		11		9	
No. years evaporation > 40.6 in.	5		5		2	

* After Birch and Andrewartha (1944).

greater in some parts of the area where *Austroicetes cruciata* lives than in others; also the difference in severity between Hawker and Hammond in 1938–39 was fortuitous, because, on the average, the risk is equally great in both these districts. This is characteristic of the sort of variability which is nearly always found when natural populations are studied thoroughly.

Most eggs which survive the summer develop normally and survive the winter. But another hazard may occur when newly hatched nymphs are ready to emerge from the egg pods. The egg pod of *Austroicetes cruciata* is closed by a "cap" of the same material that forms its walls. The female also rakes loose soil over the top of the pod, covering it to a depth of about 1 mm. The soil becomes compact during winter. If, at the time when the nymphs are ready to emerge, the surface of the soil is dry, the cap may be cemented on so firmly that the nymphs cannot push it off. We sometimes found nymphs which had been trapped in this way and had died from starvation, but we never found any extensive situation where the death-rate from this cause exceeded a few per cent. Nevertheless, it is conceivable that a drought coinciding with the main hatching of the nymphs might cause many of them to be trapped in the egg pods. So we estimated the probability of such an event. Experiments in the laboratory showed that newly emerged nymphs could live for about 5 days without food. Experiments in nature showed that 0.02 inch of rain would soften the soil sufficiently to allow the nymphs to break out from the pod. Using methods which were described by Birch and Andrewartha (1944) and Andrewartha (1944b) and taking the 50 years between 1891 and 1940, we estimated the frequency with which the main hatching of *Austroicetes cruciata* would have coincided with a 6-day drought. This was found to be about 0.012, indicating that a high death-rate from this cause might happen about 12 times in a thousand years. The nymphs encounter much greater risks than this from other sources as they become older.

When the nymphs first hatch, they are not likely to be short of food because this is the season of the year when the grass is growing most strongly. During the period 1935 to 1939 when swarms of *Austroicetes cruciata* were present, the later stages sometimes experienced a shortage of food, not because there was insufficient grass in the area but because it dried up before the grasshoppers could complete their development. The nymphs did not usually suffer much from shortage of food, but the duration of life and the fecundity of the adults were largely determined by the date when the grasses dried up. As the grasses withered, the grasshoppers died from starvation. But the sheep which shared the pastures with the grasshoppers could thrive on dry grass; and it was the aim of the farmer to have standing in his paddocks enough "dry feed" to carry his flocks through the summer. Perhaps the surest evidence that the grasshoppers on no occasion between 1935 and 1939 ate more than a small proportion of their potential stock of food lies in the records of the numbers of sheep living on the same pastures during this period; the number of sheep increased from 469 thousand in 1935 to 712 thousand in 1939.

It was fortunate for our study that 1940 was a year of severe drought, with a heavy death-rate among nymphs during the spring. By measuring the severity of this drought and consulting past meteorological records, we were able to estimate the frequency with which nymphs might suffer a heavy death-rate from drought.

Many eggs were laid in the spring of 1939, and everything pointed to a generation in 1940 with probably even greater numbers. But little rain fell during the growing season of 1940, and when the nymphs hatched during the first week in September, there were already signs that food would be scarce later on. The dramatic ending to the five years of plagues was described by Birch and Andrewartha (1941, p. 95):

> In 1940 the nymphs emerged mostly during the first week of September. There was very little food available, due to the lack of rain during winter. For example, at Orroroo the total rain from April to August was 2.44 inches (mean for this period 6.09 inches). Annuals were scarce or absent from most situations; speargrass had only a few short green shoots near the base. During September and October rain occurred as light falls widely spaced. Temperature and atmospheric saturation deficit were above average. An examination of records for 15 representative towns showed that no rain fell between 21st August and 12th September; on that date between 11 and 68 points were recorded; light falls varying from 1 to 30 points were registered on 20th September; on 28th September falls between 3 and 40 points were recorded; falls ranging between 4 and 34 points were recorded on 15th October, and further falls varying from 1 point to 17 points on 29th October. No further rain was recorded until 9th November. These falls of rain were insufficient to maintain growth of speargrass and the grasshopper numbers were reduced as a result of starvation.
>
> Survey trips were made on 3rd September, 16th September, 3rd October, 29th October and 18th November. On the 3rd September hoppers were present in plague numbers in most situations examined; in one or two local situations there was evidence that death from starvation had already occurred on a large scale. On the 3rd October the hoppers were found in plague numbers in only a few favoured local situations; it was noticed that their development had been unusually slow. By 29th October the grasshoppers had practically disappeared; only two small situations were found where they were numerous. By 18th November they had disappeared almost completely. During an extensive survey which covered many districts no situation was found where there was more than an occasional grasshopper. In some places the grasshoppers had disappeared from areas close to wheat crops which were still green. Observations by many local residents provided the explanation for this. As the grasshoppers had died out over the countryside at large, the birds, which normally prey upon them, had been forced to seek out the grasshoppers in the local situations where the latter still survived. The congregation of starlings, crows and other birds had been so great that in most places the grasshoppers had been almost completely exterminated.

From this account it is clear that the primary cause of the calamity was lack of rain during winter and spring. We analyzed the records for rainfall, temperature, and atmospheric humidity for the 50 years 1891–1940 and found that droughts as severe as this one had occurred 7 times at Hammond, 5 times at Hawker, and 3 times at Orroroo. The probability of a high death-rate among grasshoppers from the three causes we have so far analyzed—a hot, dry summer, drought at the time of hatching, and drought during the nymphal stages—has been summarized in Table 1.01.

We next consider a fourth possible cause of death. The known variability of the rainfall in this area makes it likely that excessive wetness may sometimes cause the deaths of many *Austroicetes cruciata* by permitting the spread of disease among them, but we did not observe this in South Australia. Andrewartha (1944*b*) described indirect evidence that this had happened in Western Australia during July, 1938, but it was not possible to estimate the probability of such an event in the area that we studied. There is no way, other than those which we have discussed, in which weather is likely to cause a high death-rate among a natural population of *A. cruciata*.

The sheep is the only other animal of any importance that requires to eat the same food as *Austroicetes cruciata*. The farmer tries to keep a reserve of "dry feed" to maintain his flock of sheep during summer, so it is unlikely that he would ever willingly allow so many sheep on the pasture that they would appreciably reduce the stock of food available to *A. cruciata*. In special cases, as in 1940, when the usual "spring renewal" of food did not take place and the total stock of food in the area was very much less than usual, the presence of the sheep may have accentuated the shortage of the food for the grasshopper. But even in these circumstances the primary cause of the shortage was the weather.

The predators of *Austroicetes cruciata* include a scelionid, a bombyliid, and a dermestid which eat the eggs and a sphecid which preys on the nymphs and adults. None was abundant, and none made any appreciable difference to the numbers of the grasshoppers. Birds of various sorts prey on the nymphs and adults. The presence of *A. cruciata* during spring may assist the birds to provide for their young, but the number of birds in the area depends less on the number of grasshoppers than on the amount of food they can find during the remaining 9 months of the year when the grasshoppers are absent. The relative influence of the birds on the number of grasshoppers increases as the grasshoppers become fewer. This is the reverse of what is expected from a "density-dependent factor."

13.125   GENERAL CONCLUSIONS

From 1935 to 1939 the grasshoppers were present in large and increasing numbers. It is likely that they would have increased more rapidly if the females in each generation had lived long enough to lay all the eggs which they are potentially able to produce. Dissections indicated that this was about 240 eggs, the equivalent of about 12 pods. In nature the food dried up, and the grasshoppers died of starvation before they could lay all their eggs. We made no precise estimate, but we think that the average would have been about 2–4 pods per female. The precise number does not matter. The important point is that the limit to fecundity was determined by the duration of the period when succulent food was present. Thus there is a sense in which it is true that fecundity was determined by the supply of food. But the stocks of food ran out, not because they were eaten by the grasshoppers but because, in the absence of rain to keep the soil moist, the grass changed its condition, becoming useless as food for the grasshoppers. In the closing stages the little green food that remained was sparsely scattered over many plants. So, in the end, the scarcity of food which overtook the grasshoppers was more of a relative scarcity (in the sense of sec. 11.12) than an absolute one. At a slightly later stage there would, of course, be virtually no green food left anywhere, but the last grasshopper had usually died of starvation before this stage had been reached. The supply of food was therefore quite unrelated to the numbers of the grasshoppers, and it cannot in any sense be construed as a "density-dependent factor."

It was shown in section 13.124 that the records of the numbers of sheep living on the same pastures provided independent evidence that, during the period when the grasshoppers were abundant, they did not eat more than a small proportion of the total herbage growing in the pastures. It was also shown in section 13.124 that none of the causes of deaths of *Austroicetes cruciata* in natural populations could be recognized as a "density-dependent factor." Weather was most important and was clearly not "density-dependent"; the influence of predatory birds, which become important in special circumstances, is in many ways the opposite to that of a "density-dependent factor."

The rate of increase *r* was not determined by "density-dependent factors"; yet the grasshoppers did not go on increasing in numbers to the limit of their resources of food. (Places in which to lay eggs and all the other spatial requirements of the grasshoppers were in great excess of the demands made on them.) The explanation in this case is obvious and simple. Calamity in the form of the drought of 1940 overtook the population long before this stage was reached. Virtually all the grasshoppers died without leaving progeny behind them; and nearly all the deaths were due to starvation. But, once again, the shortage of food did not depend on the numbers of grasshoppers. The food was in short supply because, from lack of moisture, the grass withered before the nymphs could complete their growth. The result would have been the same, had there been few or many grasshoppers present at the beginning.

The grasshoppers died out completely from most local situations during the calamity of 1940, but they did not become extinct from the area as a whole; a few survived in isolated places here and there. It was shown in section 13.124 that similar hazards recur with certain frequencies which can be estimated from the meteorological records. This has been going on for a very long time, and the continued persistence of *Austroicetes cruciata* in this area shows that the probability of complete extinction during any one vicissitude must be exceedingly small. It has been argued that such small probabilities cannot be explained except in terms of "density-dependent factors." We think that this view fails to allow for the great heterogeneity of places where *A. cruciata* may live in this area and for the chance that a few individuals here and there will have the good luck to find themselves in places where life can be sustained. This may be a little more likely to happen if the population is large at the time that catastrophe strikes, but this does not make the weather a "density-dependent factor" in the usually accepted sense of this phrase. Indeed, the number of survivors is likely to be determined largely by the interaction of the two components of environment which we have called "weather" and "place to live"; the number of predatory birds may also be important; this does not depend on the number of grasshoppers but on the amount of food which the birds can find in the area during the 9 months when the grasshoppers are absent. We showed in section 1.1 how, as the climate becomes increasingly severe, the probability of complete extinction from a substantial area increases. Indeed, this is what happens in the areas which we regard as being outside the distribution of *A. cruciata*.

The area which we had mapped (secs. 13.123; Fig. 13.07) might be taken to represent the distribution of *Austroicetes cruciata*, or at least that part of it where they have a good chance of becoming numerous from time to time. If we had been right in concluding that the abundance of *A. cruciata* is determined largely by weather, then it should be possible to define a particular climatic zone such that its boundaries are closely related to the boundaries of the distribution of the grasshopper. In seeking to test this hypothesis, we were guided by the information given in section 13.124.

In southern Australia the climate becomes more arid as the distance from the coast increases. Therefore, the limit to the northern (inland) distribution of the grasshopper might be largely determined by the severity and frequency of drought, especially during winter and spring, when rain is needed to insure food for the active stages of the grasshopper. The mean monthly ratio of rainfall to evaporation (P/E) has often been used as an index of climate. The value of $P/E = 0.5$ has been used to estimate the duration of the growing season for crops and pastures in general (Davidson, 1936c; Fig. 11.01). Since we were concerned with especially hardy grasses, we chose to use the value $P/E = 0.25$. Since the most dangerous month for the nymphs is October (if the nymphs survive October, the adults are likely to find enough food to lay at least a few eggs), we chose to plot on the map the isocline for $P/E = 0.25$ for October. Fortunately, this isocline had been calculated and mapped (though not published) by Davidson some years earlier. So we had an independent estimate of the climate with which to test our hypothesis. The agreement in Figure 13.07 between the climatic line and the northern (inland) boundary of the distribution of *A. cruciata* is remarkably close.

The southern (coastward) limit to the distribution of *Austoicetes cruciata* is likely to be determined by the risk of high humidity, favoring the spread of disease. Since the active stages are present during spring when the weather is becoming drier, the greatest risk of encountering a spell of dangerously wet weather occurs early rather than later in the lives of the nymphs. The usual month for hatching is September, so we chose this month. We had available from Davidson's working drawings the isocline for $P/E = 1.0$, which, being double the value usually chosen to indicate sufficient moisture to maintain plant growth, seemed a suitable index for our purpose. In Figure 13.07 the isocline for $P/E = 1.0$ for September follows the southern boundary of the distribution of *A. cruciata* quite closely. The hypothesis has therefore been supported, and we considered this good supporting evidence for our general conclusion that the distribution and abundance of *A. cruciata* are determined largely by weather; there is no evidence for "density-dependent factors."

In Figure 13.07 there are quite large areas between the two climatic lines from which *Austroicetes cruciata* was absent or where local populations were found here and there in isolated situations. In some quite extensive areas the soil was unsuitable for oviposition; in others the persistence of the original vegetation has resulted in a lack of suitable plants for *Austroicetes* to eat. This was explained by Andrewartha (1944*b*). There is no need to go into the matter here.

13.13. *The Distribution and Abundance of* Porosagrotis orthogonia *in North America*

The ecology of the pale western cutworm *Porosagrotis orthogonia* in the Great Plains region of western North America was studied by Parker, Strand, and Seamans (1921), Seamans (1923), and Cook (1924, 1926, 1930). This species is common along the eastern edge of the Rockies, and isolated specimens have also been taken in California, Nevada, Arizona, and New Mexico. Cook was able to study the influence of weather on the rate of increase of *P. orthogonia* during a great outbreak which began about 1917, reached a climax about 1920, and declined during 1921 and 1922. He mapped the limits of the area where outbreaks of this species were likely to occur and he estimated the probable frequency of outbreaks.

13.131   THE BIOLOGY OF *Porosagrotis orthogonia*

The adults of *Porosagrotis orthogonia* fly during the autumn, and the females lay, on the average, about 300 eggs each. There is a faint diapause in the egg stage which is nevertheless sufficient to insure that the hatching of the larvae will nearly always be delayed until the spring. The larvae live in the soil and feed on the roots and underground stems of wheat and other plants. They complete their development by about the end of June and emerge as moths during August and September.

In seeking a place to lay eggs, the females show a strong preference for fields of stubble, in which the soil is loose at least in places; eggs were often laid in the hoofprints left in soil by horses. The newly hatched larvae live on the surface of the soil for perhaps 2 weeks. Then they burrow into the soil and remain below the surface for the rest of their lives except when they are driven above ground by excessive wetness. The following quotation from Cook (1930, p. 11) shows how sensitive they are to the moisture content of the soil:

> Many examinations of infested fields have led to the conclusion that the depth at which they feed is regulated by the moisture in the soil. In May and June in Montana the rainfall is very irregular. Following a shower the surface soil will dry out and crumble, leaving a very definite depth at which the soil is still moist. Farmers call this depth the "moisture line" and use its position as an indicator of the amount of moisture still available for the wheat. In extremely dry seasons the line may lie as much as 6 inches below the surface, but it is usually found within 3 inches of the surface. As the moisture line is the place of highest evaporation, it is a cool place, definitely cooler than the soil half an inch above it. Careful study has shown that *orthogonia* larvae will usually be found just above the moisture line while they are active. . . .
>
> In wet weather the moisture line comes to the surface, and so do the larvae. Several times larvae have been observed wandering around on the surface in bright sunlight just after a heavy shower. As soon as the excessive moisture drains from the surface layer, the larvae dig into the soil and again follow the moisture line.

Cook did not measure fluctuations in the birth-rates of natural populations. It is likely that fluctuations in the value of r would be determined largely by fluctuations in death-rates and that variation in the birth-rate would be relatively unimportant.

13.132   CAUSES OF DEATHS AMONG NATURAL POPULATIONS OF *Porosagrotis orthogonia*

Cook made certain observations and did several experiments which showed that larvae in soil that was uniformly friable moved at random, except when very close to the stem of a wheat plant or some other food. But wherever the soil was unevenly compact, the larvae moved along the line of least resistance. In a field of wheat which had been sown with a drill, they moved along the lines made by the drill. As a result of this behavior, larvae that are living in a field of wheat find food quite readily. But in uncultivated land, even if the vegetation is equally dense, their chance of finding food may be less. That is, they may suffer from a relative shortage of food in the sense described in section 11.12.

A quotation from Cook (1930, p. 50) emphasized this point:

> Before 1911 *P. orthogonia* was a rare insect. There are a few records of the capture of the moths but no records of damage. From a study of its present habits, it seems safe to assume that in former times it was confined to places where the soil was very loose and dry. Such situations are found in the lighter soils of Montana on knolls and along the edges of benches. It is evident that such situations were not large in area, so that the total area suitable for this species must have been small. Only in these places was it able to feed underground, and, if it occupied other places, where it must feed at the surface, it would be subject to the same parasites and enemies as our other common cutworms, and be kept to small numbers by them. With the extension of dry-land wheat-growing the situation was changed. Not only was the area of loose soil enormously increased, but the soil was planted with a tender and succulent plant. This enabled the species to increase in numbers, and gradually to accumulate a normal population so large that it formed a nucleus for a population so large that it would cause severe damage when conditions became favourable. In order to reduce it permanently to its former status of rarity, the cultivation of grain would have to be discontinued.

The presence of extensive areas of wheat certainly permitted *Porosagrotis orthogonia* to achieve relatively enormous numbers during the outbreak of 1917–21. In certain local situations there were too many for the supplies of food, and many larvae died from starvation or were eaten by their fellows. Nevertheless, they did not, even during their period of greatest abundance, destroy more than 45 per cent of the wheat in the area studied by Cook. So the decline, when it came in 1921 and 1922, was not due to any substantial or widespread shortage of food.

Seamans (1923) observed that the numbers of *Porosagrotis orthogonia* were usually depleted during a generation which was forced to the surface frequently by wet weather; and he postulated that *r* would remain negative (or, if positive, it would be low) for any generation which was exposed to more than 10 wet days (0.25 inch or more per day) during May and June. He attributed the high death-rate on these occasions to predation by a variety of insects and several birds, which, he argued, would kill many more caterpillars if they were exposed on the surface than if they were hidden in the soil. But Cook studied certain populations at the time when the outbreak was declining in 1922; the death-rate exceeded 90 per cent, and Cook found that more than half the deaths were due to disease. Having listed the names of 10 insects and 3 birds which preyed on the larvae of *P. orthogonia*, he concluded: "In spite of this list of enemies, it must be said that they were apparently of small value in terminating the outbreak of *orthogonia* in 1922. In that year, when the population of cutworms was already greatly reduced, the combined efforts of parasites and predators accounted for about 50 to 60 per cent of the remaining population, near Havre, but there is no indication that their work did any more than slightly aid climatic factors and disease in terminating the outbreak."

13.133   THE INFLUENCE OF WEATHER ON THE DISTRIBUTION AND ABUNDANCE OF *Porosagrotis orthogonia*

Cook did not measure the absolute numbers of *Porosagrotis orthogonia*. But he collected, for a large number of farms, records which showed, for each generation, whether it was larger or smaller than the one which had preceded it. He did this for the five years 1919–23 for a large number of districts. He then compared the monthly records of rainfall and temperature with these data and demonstrated a close relationship between the rainfall during May, June, and July and the rate of increase of *P. orthogonia*. The rate of increase was determined largely by the survival-rate, which was likely to be high when the monthly rainfall for May, June, and July averaged less than 1.3 inches, and likely to be low if the monthly rainfall during this period exceeded 1.7 inches.

A hypothetical climate, based on the results of this study, was arbitrarily chosen as being optimal for the multiplication of *Porosagrotis orthogonia*. The winter was dry, with about ½ inch of rain per month; the spring was damp, with April, May, and June averaging about 1.5 inches per month; the summer was dry, with about 4 inches falling during July–October. The temperature ranged from about −5° C. in December to about 20° C. in July. The records for monthly rainfall and temperature for several hundred stations in western United States were then analyzed in the same way and compared with this hypothetical climate.

In this way Cook delimited a zone which included nearly the whole of

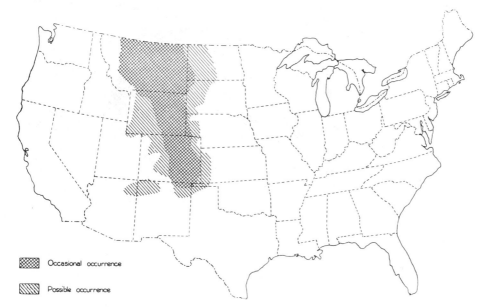

Occasional occurrence

Possible occurrence

FIG. 13.08.—The area in the United States where *Porosagrotis orthogonia* might be expected to maintain dense populations from time to time. (After Cook, 1930.)

Montana and extended southward through Wyoming and Colorado. He postulated that, inside this zone, *Porosagrotis orthogonia* would be a common species, occasionally multiplying to numbers that constituted a plague. He mapped another narrower zone bordering the first, where, he postulated, *P. orthogonia* would be absent or present in lower numbers but where it might much less frequently multiply to great numbers. These zones are shown in Figure 13.08. He tested his hypothesis by mapping all the records of all known plagues of *P. orthogonia*. Without exception, they fell inside the limits of the zone which he had mapped entirely by reference to meteorological records.

Cook also estimated the frequency of outbreaks of *Porosagrotis orthogonia* in Montana. We give his own account of his calculations:

A study of the Montana outbreaks indicates that one favourable year may increase the number of cutworms sufficiently to cause slight damage and local outbreaks, but two successive favourable years are necessary to produce a severe and widespread outbreak. Curves were fitted to the rainfall distributions for May, June and July for [a number of stations]. . . . It was found that single years with less than 4 inches of rainfall from May 1 to July 31 should occur, on the average, once in 4 years near Helena, once in 5 years near Havre and Crow Agency, once in 6 years near Miles City, once in 8 to 9 years near Poplar and Glendive and less than once in 10 years at Bozeman. These figures indicate the probability of mild outbreaks of a local character [Fig. 13.09].

In order to determine the probability of two successive dry years, it was necessary to form a frequency curve for the differences in rainfall between successive years and compute the probability of a variation of less than one inch in successive years. This probability, multi-

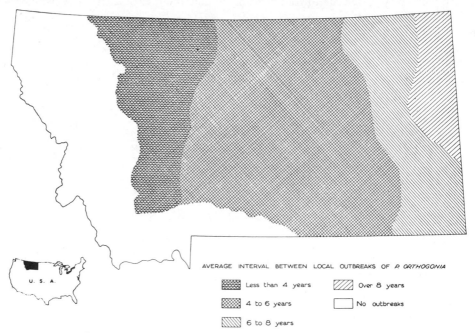

Fig. 13.09.—Areas in Montana where local outbreaks of *Porosagrotis orthogonia* might be expected to develop with the frequencies indicated by the shading. (After Cook, 1930.)

plied by the probability of one favourable year, gave the probability of two successive dry years. Two successive dry years may be expected about once in 16 years near Helena, once in 23 years near Havre and Crow Agency, once in 30 years near Miles City, once in 40 years near Bozeman and once in 60 or 70 years near Poplar and Glendive. These figures indicate that severe and widespread outbreaks may be expected only at long intervals in most parts of Montana [Fig. 13.10].

13.134   GENERAL CONCLUSIONS

The amount of readily available food and the area of friable soil which is suitable as a place where *Porosagrotis orthogonia* may live have been increased enormously as the country was developed for agriculture. The numbers of *P. orthogonia* have increased as a consequence. Not only do they become much more numerous during periods of maximal abundance, but they also maintain a larger population during periods of minimal abundance.

Nevertheless, *Porosagrotis orthogonia* does not fully occupy all the suitable places to live or consume all the available food, even when the caterpillars are most numerous, far less so when they are scarce. The two components of environment which we have called "food" and "place to live" are never in short supply, at least in the absolute sense; there is no suggestion of any "competition" for these resources, except perhaps locally during an outbreak; and there-

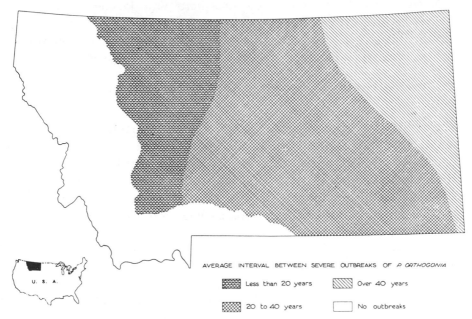

AVERAGE INTERVAL BETWEEN SEVERE OUTBREAKS OF *P. ORTHOGONIA*

U. S. A.

Less than 20 years   Over 40 years

20 to 40 years   No outbreaks

FIG. 13.10.—Areas in Montana where severe and widespread outbreaks of *Porosagrotis orthogonia* might be expected with the frequencies indicated by the shading. (After Cook, 1930.)

fore it is not necessary to consider either "food" or "place to live" in looking for a "density-dependent factor" in the environment of *P. orthogonia*.

Although these two components had changed greatly in the past, they had become relatively stable by the time of this investigation. With "food" and "place to live" not varying much, the numbers attained during periods of relative abundance depended on the weather, especially on the duration of favorable periods. A run of several consecutive favorable years might result in enormous numbers, like those observed during 1920. But the probability of such a run is low. Usually the period of increase would be expected to change to one of decrease before the population had become so dense as this.

Cook showed how catastrophe in the form of excessive wetness at a critical time may overtake a flourishing population and cause it to decline with dramatic suddenness. He indicated the probable frequency of such catastrophes and the depth to which the numbers might decline after an outbreak. He did not investigate how the numbers were determined during periods when the insects were few and decreasing. This is always a most difficult part of any investigation. It is true that *Porosagrotis orthogonia* may be eaten by several species of insects and birds; but in 1922, after the outbreak had been in progress for 5 years, the combined activities of all predators destroyed less than half the population. It is therefore unlikely that predators would have much influence on the numbers of *P. orthogonia* when they were already few. We do not know the answer to the question: Why does not *Porosagrotis orthogonia* become extinct in the Great Plains region? But we can say with assurance that this investigation provided no evidence that the right answer might be: Because its numbers are regulated by "density-dependent factors."

13.14   *Fluctuations in the Numbers of* Choristoneura fumiferana *in Canadian Forests*

The spruce budworm *Choristoneura fumiferana* is indigenous in North America. From time to time, outbreaks have occurred which have endured for perhaps 5 or more years and, in some instances, extended over thousands of square miles of forest. The caterpillars eat the buds, the flowers, and the leaves of spruce, balsam fir, and other coniferous trees. When they are numerous, the budworms may defoliate the trees, doing enormous damage, especially to spruce. After an outbreak has subsided, many years may pass before the budworms become numerous in the same area again. There is evidence, based especially on the study of growth-rings, that the numbers of this indigenous insect have been fluctuating like this for centuries.

Outbreaks may be associated with weather of a particular sort, and this relationship has been studied especially by Wellington *et al.* (1950) and Wellington (1952). The following brief summary is taken chiefly from these two papers. There are numerous other relevant publications dealing especially with the biology of *Choristoneura fumiferana*, and most of these are quoted in the two papers which we have cited. Wellington's studies on the behavior and physiology of *C. fumiferana* have been summarized in sections 7.12, 7.16, 8.31, and 8.32.

A firm obligate diapause occurs in the first larval instar of *Choristoneura fumiferana*. The larvae, hatching during the autumn, do not feed but immediately seek winter quarters, where they spin a "hibernaculum," in which they remain until the following spring. The active stages are present during May, June, and July. There is, because of diapause, only one generation a year, and the same stage in the life-cycle recurs at the same season each year.

The survival-rate among the hibernating larvae is likely to be great when snow falls early and the temperature remains consistently low. Conversely, the death-rate may be high when fluctuations in temperature cause repeated thawing and freezing. The survival-rate among the active stages is greater, the greater the abundance of staminate flowers and the greater the insolation of the places where the larvae are living. Balsam firs that are approaching maturity tend to produce staminate flowers in abundance. Stands of spruce and balsam fir in which the trees are approaching maturity allow the penetration of more light and radiant heat from the sun. So the value of *r* for *Christoneura fumiferana* is likely to be higher in stands of mature balsam firs or in mixed stands of spruce and balsam firs in which most trees are approaching maturity. No matter how favorable the other circumstances may be, there is little likelihood of *r* being large in stands where the trees are immature (for this is not the most suitable food) or where the conifers are shaded by mature poplars and birches (for the temperature is too low in such places).

Thus, after the destruction of the mature trees of spruce and balsam fir by the budworms or by fire, which may follow severe defoliation by the caterpillars, or by logging, there is a period during which the forest will support only a sparse population of *C. fumiferana*, no matter how favorable the weather may be. Later, when the coniferous trees on which the budworms feed are approaching maturity, the insects may increase in numbers if other circumstances favor them. There is no evidence that this depends to any important extent on predators or any sort of "density-dependent factor." This point was emphasized by Wellington *et al.* (1950, p. 329): "A native insect that has this indicated background and that exhibits violent fluctuations in population density cannot be governed by parasites and diseases during its periods of minimal or initially increasing density so much as by a combination of forest type and climate. The current survey has indicated that endemic populations in the susceptible foci begin to grow to outbreak proportions when the climatic control is relaxed. There has been no indication of control or release of populations by parasitic or pathogenic agents during the long period of minimal density or during the initial population growth in the foci."

The active stages of the budworm are favored by sunny, cloudless weather with little rain during June and July. Figure 13.11 shows how closely the activity of the caterpillars was associated with the rainfall during June. Larvae move away from places that are too wet (secs. 7.12, 7.16). This interrupts their feeding, and many of the larvae die. The hibernating larvae are also favored by keen cloudless weather during the winter. This sort of weather, during both summer and winter, is associated with clear, dry air which is usually found in what the meteorologists call polar-maritime or polar-continental air masses. These names refer to the regions where the air masses originated and acquired their characteristic qualities. The other sort of air mass which commonly occurs in the boreal region of Canada is called "tropical-maritime." This is likely to be associated with warm, humid weather in the winter and rains and cloud during summer; such weather does not suit *Choristoneura fumiferana*.

Centers of low pressure (cyclones) occur relatively infrequently when the weather is being dominated by polar-continental or polar-maritime air masses. So the frequency of cyclones provides an indirect measure of the relative frequencies of the two sorts of air mass and hence of the weather. The meteorological services of Canada maintain records going back many years, from which Wellington was able to calculate the annual frequencies of cyclones in different parts of the country. Figure 13.12 shows the annual number of cyclones for each of 8 districts. The analysis covered 20 years for each district,

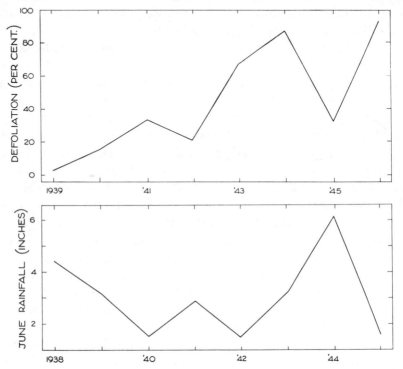

F<small>IG</small>. 13.11.—The upper curve shows the extent to which *Choristoneura fumiferana* defoliated certain trees of balsam fir each year between 1939 and 1946. The lower curve shows the amount of rainfall during June each year between 1938 and 1945. Note the shift of 1 year in the abscissae. (After Wellington *et. al.*, 1950.)

starting in each case 14 years before the beginning of an outbreak of *Choristoneura fumiferana*. Since the outbreaks occurred on different dates in each district, the curves, though they appear in the one diagram, cannot be related to a common date.

The dates of the beginnings of the outbreaks are indicated by arrowheads in Figure 13.12. In some instances the date on which the outbreak "began" was fixed by direct observation of severe defoliation; but in most instances it was inferred from a study of growth-rings. Wellington concluded from these results that outbreaks have usually begun during periods when the weather was being dominated by polar air masses, which would probably result in below-average rain and cloud and above-average hours of sunlight during the summer and consistent cold with a continuous covering of snow during the winter.

After an outbreak has been in progress for several years, predators and diseases become relatively more abundant. But, as the following quotation from Wellington *et al.* (1950, p. 329) shows, it is doubtful whether they are of

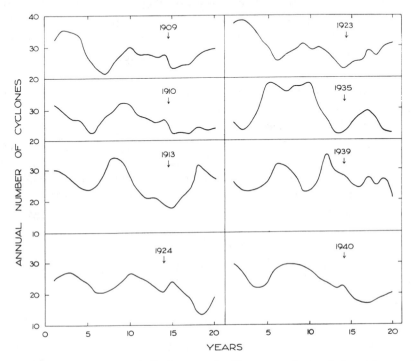

Fɪɢ. 13.12.—Major outbreaks of *Choristoneura fumiferana* were known to be under way in Canadian forests, in one place or another, in 1909, 1910, 1913, 1923, 1924, 1935, 1939, and 1940. Each curve shows the number of cyclones recorded in the particular area during the 15 years before and the 5 years after these dates. For further explanation see text. (After Wellington, 1952.)

primary importance in causing the outbreak to decline: "During the first portion of the outbreak period, biological control agents increase in numbers; but observations in the Lake Nipigon infestation indicate that the spruce budworm population begins to experience other troubles, some of which, like starvation, are self-inflicted, shortly before the biological agents assume an important role. The relationships are complex but there are now field indications that starvation and specific reactions to meteorological factors that drive large numbers of larvae from the trees prior to pupation may decrease the spruce budworm population just enough to make it extremely vulnerable to the rapidly increasing populations of biological agents."

13.2   OF FOUR NATURAL POPULATIONS IN WHICH THE NUMBERS ARE DE-
TERMINED LARGELY BY "OTHER ORGANISMS OF DIFFERENT KINDS"

It is well known for many species of invertebrates that their numbers are
determined largely by predators (secs. 10.321, 12.23). The circumstances in
which the predators are likely to keep the prey scarce are fairly well understood: briefly, the numbers of the prey are likely to be fewer, the more the
predator excels the prey in its capacity for multiplication, dispersal, and
searching; these qualities are relative and may change with circumstances—
for example, weather or the sorts of places available to the prey to live in.
This generalization may be made with assurance, at least with respect to insects, because much work has been done by entomologists interested in the
"biological control" of pests. The economic aspect has been emphasized. There
has been a tendency to classify predators according to whether or not they
exercise "economic control" over the pest. The "unsuccessful" ones have received little attention; the "successful" ones have afforded better opportunities
for observation, and more has been written about them. Nevertheless, we do
not know of a natural population in which the numbers are largely determined
by predators that has been the subject of a sustained quantitative investigation
comparable with those mentioned in section 13.1.

The importance to be attached to predators of vertebrates is more doubtful.
With species like the muskrat and the bobwhite quail, which show a keen
awareness of "territory," it seems fairly clear that predators are not so important as the "carrying capacity" of the area where the animals are living
(secs. 10.322, 12.31, 13.34). With the larger ungulates and perhaps also with
some of the lower orders, predators may be more important. But we do not
give any examples of this.

In this section we have, in Lloyd and Reynoldson's studies of the ecology
of certain animals which live in sewage-filter beds, two good examples of
natural populations in which the numbers are largely determined by non-
predators which also live in the sewage beds. We have an instructive analysis
by Ullyett of the causes of deaths in local populations of *Plutella maculipennis;*
this reveals an important interaction between predators, diseases, and weather.
And we have in *Rhagoletis* a nicely worked-out example of an interaction between predators and place to live in. The activities of a nonpredatory animal
(man) result in a large increase in the number of places for *Rhagoletis* to live in.
The outcome is that a predator which preys on *Rhagoletis* finds itself with a
shortage of food in the relative sense described in section 11.12. But we do
not have a well-documented study of a natural population in which the numbers are determined chiefly by predators.

13.21 *The Numbers of* Enchytraeus albidus *in a Sewage-Filter Bed near Huddersfield*

The ecology of the oligochaete worm *Enchytraeus albidus* was discussed by Reynoldson (1947*a*, *b*, 1948) and by Lloyd (1943); the last paper is about the fly *Psychoda alternata*, which is the only other animal of any importance in the environment of *E. albidus*.

The sewage beds at Huddersfield differ from others which were studied (see sec. 13.22), in that the sewage contains an unusually large proportion of "chemical waste" from factories. It is so toxic that the large variety of insects and worms usually found in sewage beds is, at Huddersfield, reduced to a few, two of which, namely, *Enchytraeus albidus* and *Psychoda alternata*, are the only ones that are common. Although *E. albidus* can survive in the sewage bed at Huddersfield, they do not attain the high numbers which are characteristic of the other species which live in other beds where the sewage is not toxic. Experiments in the laboratory showed that the sewage from Huddersfield was toxic in some degree even to *E. albidus*, especially to the young stages, but it was much more toxic to *Lumbricillus lineatus*, which is common in other beds.

The sewage was sprayed onto the surface of the bed, which consists of a layer of "clinkers" 6 feet deep. It percolated down, becoming less toxic as it penetrated more deeply, and eventually the effluent flowed away at the bottom. About 11 million gallons were applied daily to 2 beds, each 200 feet in diameter.

A coating of fungi, bacteria, algae, and inert organic matter formed over the surface of the clinker. This was the chief source of food for both *Enchytraeus albidus* and *Psychoda alternata*. The feeding of the latter aerated the mass; in their absence, anaerobic putrefaction would make the bed quite unsuitable for *E. albidus*, and none could live there. Nearly all the larvae of *P. alternata* were found in the top 12 inches, whereas most of the worms occurred below this depth, so there was no question of the one eating the food that the other required. Nevertheless, in another, much more subtle, way the activities of *P. alternata* were largely the cause of the prevailing low numbers of *E. albidus* in the beds at Huddersfield.

The feeding of a lot of larvae of *Psychoda alternata* caused the coating of fungi, algae, etc., to slough away from the clinker. It was then washed down through the beds and out with the effluent. The worms, partly because they have the habit of clinging to small loose particles of solid, were washed out of the bed in great quantities. Reynoldson sampled the effluent during the spring (when this phenomenon was most pronounced) and estimated that there were about 19,000 worms and 28,000 cocoons containing 100,000 eggs being washed out of the bed daily. And he commented: "Losses much less than these would be a severe drain on such a small population."

If the worms were exposed to this risk continuously, one would expect the population to become extinct. But invariably the washing-out (or "off-loading," as it is called technically) ceased for a period, during which the clinker became covered with a fresh coating of fungi and algae. The worms multiplied during this period. At the same time, the larvae of *Psychoda alternata* had pupated and later emerged as adults. They mated and laid eggs, which in due course gave rise to another batch of feeding larvae. These, in turn, caused another washing-out or "off-loading" to take place. The worms were once again washed out of the bed, and again their numbers were greatly reduced. The rate of increase of *Enchytraeus albidus* between these periodic disasters was slow, perhaps because of the toxins in the sewage or for other reasons that are not understood, so the worms never attained numbers that are comparable with those attained by other species in beds where the sewage is not toxic. On the other hand, they never became extinct, because the phenomenon of off-loading was temporary and ceased long before the last worm had been washed out.

The periodic recurrence of off-loading was determined by the periodic fluctuations in the numbers of *Psychoda alternata*. Lloyd (1943) studied the ecology of this species and showed that the fluctuations were determined partly by the progression of the seasons and partly by fluctuations in the supply of food brought about by the excessive numbers of the larvae themselves. A similar phenomenon was described in greater detail by Nicholson (1950) for an experimental population of *Lucilia cuprina*.

Figure 13.13 illustrates for one year the trends in the numbers of *Enchytraeus albidus* and *Psychoda alternata*. The worms bred most actively during the cooler parts of the year, especially during September–October and again during March–April. The flies developed slowly during winter; they began to be numerous about May and continued so throughout summer. The numbers of adults emerging from the beds fluctuated widely, with a period of about a month between maxima. There was a corresponding period in the fluctuations of the larvae feeding in the bed. The same phenomenon continued during the winter, but the numbers were lower. Off-loading occurred at a reduced rate and less frequently during the winter, and this enabled the worms to multiply at this season. They reached their greatest abundance during late winter and early spring. They were reduced to low numbers by the massive off-loading which occurred during the spring and did not recover until the next winter. There is no need to invoke the concepts of "density-dependent factors" or "competition" to explain how the numbers in this natural population of *E. albidus* are determined.

13.22   *The Numbers of* Metriocnemus hirticollis *and* Spaniotoma minima *in a Sewage-Filter Bed near Leeds*

Lloyd, Graham, and Reynoldson (1940) listed 3 species of Oligochaeta, 2 species of Mollusca, and 7 species of Diptera which are relatively common in the sewage-filter beds at Knostrop near Leeds, and they discussed the ecology of some of them. Other papers have been published on the same subject, and we have referred especially to those by Lloyd (1937, 1941, 1943).

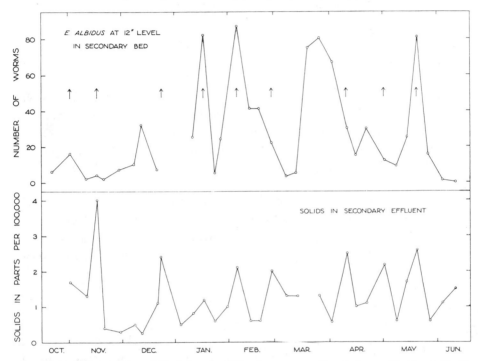

Fig. 13.13.—The number of *Enchytraeus albidus* in a sewage-filter bed near Huddersfield. The lower curve shows the fluctuations in the amount of solids washed out of the bed. The arrows indicate maxima in the process of off-loading. (After Reynoldson, 1948.)

Each bed was essentially a large heap of water-worn gravel, 6 feet high, contained by rectangular walls and provided with a drainage outlet at the bottom. The sewage was sprayed on the surface and allowed to trickle over the stones. In the top foot or so of the bed a dense growth of fungi and algae covered the surface of the stones. This growth, together with solids from the sewage which were trapped in it, supplied a rich and, at certain seasons of the year, abundant source of food for those animals which could live as near the surface as this. The chief food for the animals living in the depths of the bed was a slimy zoögloea which coated the stones in these regions.

Despite the almost continuous application of sewage of constant composi-
tion at about the same rate throughout the year, there were marked fluctuations in the amount of food available to the animals living in the bed. This was
partly because the growth and senescence of the algae and fungi were influenced by the progression of the seasons and partly because, as we saw in
section 13.21, this material had a tendency to slough off the stones and be
washed out of the bed when the concentration of animals feeding in it became
great.

During the winter the crust of algae and fungi accumulated on the stones,
which became heavily incrusted with it. The thicker this covering became, the
more likely it was to be loosened by frost and to flake off in big lumps. Such
large fragments would sink into the bed and remain there as food. On the other
hand, when there were many worms and larvae feeding in this growth, it was
more likely to slough off in small fragments, which were mostly washed through
the bed and passed out with effluent, carrying many worms and insects with
them. This "off-loading" not only caused a periodic shortage of food in the bed
but also depleted the populations of certain species by carrying many of them
out of the bed. Off-loading went on continuously, but it reached a peak during
the spring.

The sewage contained little "chemical waste," and, as a consequence, this
bed harbored many more species than did the bed at Huddersfield (sec. 13.21),
and their numbers were, on the whole, greater than those of the species which
inhabited the bed at Huddersfield. Some species were found only or mostly
near the surface; others were found only in the depths; and there were still
others which occurred throughout the bed.

The different species were favored by different ranges of temperature, and
they also responded differently to other components of the weather. With each
one its chance to survive and multiply was influenced by the activities of other
species living in the bed. None of these was an obligate predator, but there were
some that would destroy and eat certain other species if they happened across
them in the course of their feeding. Except for strict predators, all the categories of other animals which we listed in section 10.01 may be recognized in
the environments of one or another of the animals living in this sewage-filter
bed; and the ecology of these species, as revealed by Lloyd and his co-workers,
is both complex and interesting.

The fly *Psychoda alternata*, which comprised over 95 per cent of all the
insects in the bed at Huddersfield, was relatively much less abundant at
Knostrop. At Huddersfield, *P. alternata* had almost sole use of the rich resources
of food in the top foot or so of the bed, because the toxic sewage did not permit
other species to live there. But at Knostrop, *P. alternata* was quite rare in the
top foot, because this layer was usually dominated by *Metriocnemus longitarsus*
and other species which destroyed the eggs and young larvae of *P. alternata*.

Most of the eggs of *P. alternata* were probably laid near the surface, but those that were laid in the depths had a better chance of surviving and probably contributed the greater proportion of adults to the population. It would be interesting to follow the ecology of these and a number of other species more closely, but space permits us to mention only two others, the chironomids *M. hirticollis* and *Spaniotoma minima*. The former was one of the less abundant species; the latter was among the most abundant ones. The annual total number of flies trapped during the 8 years from 1934 to 1941 varied between 76 and 846 for *M. hirticollis* and between 11,000 and 38,000 for *S. minima*. These figures are based on the weekly counts of the numbers of adults emerging from the bed into small traps placed on the surface of the bed.

Generations of *Metriocnemus hirticollis* overlapped, and adults were taken in the traps during every month of the year. They were relatively scarce during autumn and winter and were usually most numerous during April–June. The adults of *M. hirticollis* require to form a swarm in order to mate. The females returned to the bed to lay eggs, but showed little tendency to penetrate deeply; as a consequence, nearly all their eggs were laid in the top foot. The full life-cycle of *M. hirticollis* required about 42 days at 20° C. and 123 days at 10°; none completed its development below 7° C. The temperature of the bed varied from about 8° C. during midwinter to about 17° C. during midsummer. Consequently, the activity of *M. hirticollis* was largely restricted to summer; eggs laid during autumn had little chance of becoming adult before June. The eggs laid during June (when the flights of adults were usually at their greatest) had ample time to become adults before winter, and one might expect the adults of *M. hirticollis* to be abundant during autumn. In fact, they were nearly always scarce at this season. Lloyd (1943) attributed this to the activities of the related chironomid *M. longitarsus*.

This species thrived at much lower temperatures than did *Metriocnemus hirticollis*. A full generation required 26 days at 20° C., 94 days at 10°, 153 days at 6.5°, and 243 days at 2° C. As a consequence, eggs laid early in the autumn emerged as adults during winter; usually circumstances were sufficiently favorable at this season, and another generation was produced before spring. This led to a large flight during May. The females laid their eggs near the surface, and the larvae established themselves in the top foot or so of the bed. Large flights of the first species, *M. hirticollis*, rarely occurred before June. So the females of *M. hirticollis* usually laid their eggs in the very zone that was already densely occupied by partly grown larvae of *M. longitarsus*. The latter are not obligate carnivores, but they readily devoured the eggs and small larvae of the former when they came across them in the course of their feeding. The consequence was that very few of the larvae of *M. hirticollis* survived, and the autumnal flights of this species were usually quite small. This sequence of events happened every year. In some years when *M. longi-*

tarsus were less numerous than usual, the rate of survival among *M. hirticollis* was higher, and the numbers of adults caught in the traps were correspondingly greater; the annual totals varied in the ratio of 1:13. It is therefore pertinent to inquire into the causes of fluctuations in the numbers of *M. longitarsus.*

Lloyd (1943) recognized two important causes for fluctuations in the numbers of *Metriocnemus longitarsus.* The most important was the weather during June, when, as a result of the large flights of adults during May, there were many eggs and young larvae near the surface. Dry weather during June allowed the surface to dry out and caused the deaths of many eggs and young larvae. On the other hand, moist weather during June insured a high rate of survival among this generation, and this, as we have seen before, correspondingly reduced the chances of survival of *M. hirticollis.*

The other cause for fluctuations in the numbers of *Metriocnemus longitarsus* was the weather during the winter: they were usually less numerous after a warm winter. Since this species was able to develop actively at moderately low temperatures, the opposite result might have been expected. Lloyd (1943) explained the anomaly by reference to the activities of the worm *Lumbricillus lineatus.* This species was most active during the winter; a generation was completed in 110 days at 10° C. and in 170 days at 7°. During a warm winter the worms might complete one generation and start another before the spring; but when the winter was colder than usual, the first generation might not be completed until quite late in the spring. Since the feeding of the worms in the crust of fungi and algae which covered the stones was the chief cause of off-loading, and since this depended largely on the concentration of worms in the surface foot or so of the bed, it is clear that off-loading would begin much earlier after a warm winter when the worms had been multiplying rapidly. This was indeed what happened. After a warm winter the stocks of food for the larvae of *Metriocnemus longitarsus* were depleted much earlier than usual, and the larvae themselves faced severer risks of being washed out of the bed. The fluctuations in the numbers of *M. longitarsus* which could be associated with this cause were considerable: the number of adults trapped during May varied from 20 in 1938 to 860 in 1941.

The numbers of *Metriocnemus longitarsus* were not great enough during any of the 8 years between 1934 and 1941 to kill all the *M. hirticollis* in the sewage-filter bed at Knostrop. If, as a result of some extraordinary circumstances, *M. hirticollis* had become extinct from this bed, it is hardly likely to have died out at the same time from all the other beds and other places where it occurs naturally. It is more likely that somewhere in the vicinity some other natural populations would have survived, which would in due course recolonize this particular bed. Nevertheless, the destruction of *M. hirticollis* by *M. longitarsus* was severe enough and consistent enough to prevent *M. hirticollis* from becoming numerous in this particular bed. It is important to notice that the numbers of *M. longitarsus* were not influenced in any way by the numbers of *M. hirticollis;* therefore, *M. longitarsus* was not acting as a "density-dependent factor" in the environment of *M. hirticollis.*

One of the most abundant insects in the sewage-filter bed at Knostrop was the small black chironomid, *Spaniotoma minima.* This species, like *Metriocnemus hirticollis,* emerged from the bed to form mating swarms, and the females returned to lay their eggs near the surface. Also, like *M. hirticollis,* the eggs and young larvae were likely to be destroyed by the older larvae of *M. longitarsus* when the two species occurred together in a densely populated medium. In fact, as with *M. hirticollis,* any egg laid after June had little chance of living long enough to become an adult. Its chance was perhaps a little higher because the larvae of *S. minima* had a tendency to move toward the depths of the bed, and consequently they spent less time in the zone where *M. longitarsus* abounded.

The life-cycle of *Spaniotoma minima* occupied 29 days at 20° C., 80 days at 10°, and 260 days at 2°. This ability to continue developing at low temperatures and the relatively short period required for a generation at moderate temperatures enabled *S. minima* to continue breeding throughout the winter and to produce a relatively large flight in the spring, usually in advance of the major flight of *Metriocnemus longitarsus.* This enabled the next generation of larvae to establish themselves before the chief danger from *M. longitarsus* had developed, and there were usually further large flights of *S. minima* during August. The progeny from these flights were largely destroyed by *M. longitarsus;* and, by autumn, the numbers of *S. minima* had usually dropped well below what they might have been in the absence of *M. longitarsus.* But the numbers of *S. minima* nevertheless remained relatively high, because their short life-cycle, their capacity to develop during winter, and their habit of colonizing the depths of the bed as well as the zone nearer the surface enabled them to recover rapidly from the setbacks imposed by *M. longitarsus.*

Lloyd discussed these results in terms of competition. This usage implies that the egg or small larvae of one species is eaten by its "competitor," while there is still no great shortage of the resource for which they are "competing." This does not seem to come within the limits of any definition of competition that is acceptable. If one must seek to use a single abstract noun to describe these complex phenomena, perhaps "predation" would be nearer the mark. But this is also misleading, because the relationships between *M. longitarsus* and the other species which are eaten are very different from those which characterize the strict predators and their prey, such as were described in section 10.32 and elsewhere. It is better to leave the phenomenon unnamed than to call it by a name that is misleading.

In the summaries given in this section we have considered briefly for both *Metriocnemus hirticollis* and *Spaniotoma minima* how the speed of development and especially the expectation of life may be influenced by the several components of environment. Temperature, moisture, and food were all of some importance, but the component of chief importance was "other animals." As usual, it was necessary to take into account the place where the animal was living in order to assess the influence of the other components on the value of r.

13.23   *The Numbers of* Rhagoletis pomonella *in Fields of Blueberries in Maine*

The larva of *Rhagoletis pomonella* lives in the fruits of the blueberry. The following account of the ecology of *R. pomonella* (Diptera, Trypetidae) is drawn from Lathrop and Nickels (1931). In order to appreciate how numbers of *R. pomonella* are determined, it is necessary to know something of how blueberry land is "farmed" in Maine.

Blueberry is the name given to several species of the genus *Vaccinium*. They are low shrubs, growing about 6–12 inches high, and, in certain circumstances, they bear heavy crops of edible berries. The species are indigenous to northeastern North America, where they originally occurred as a sparse undergrowth among tall shrubs or along the margins of forests. When the tall shrubs and the trees were destroyed, the blueberries invaded the areas where the forests had been. It was long ago discovered that if this vegetation were subjected to periodic "burnings," the blueberries would become dominant. Lathrop and Nickels (1931, p. 262) described the procedure:

On bright, calm days in early spring, after the snow leaves, but before the frost starts out of the ground, and while the blueberry plants are still thoroughly dormant, the surface litter is ignited and under favourable conditions the fire sweeps the land clear of vegetation. . . . The above ground parts of the blueberry plants are removed by the process of burning. The root system is unharmed by the fire, and the plant responds by a greatly accelerated vegetative growth during the summer immediately following the burn, but the plant produces no fruit during this first season. During the second summer a large crop of berries is produced. After this first, abundant crop the production of fruit decreases each season until the yield becomes practically nothing; unless the land is again burnt-over to rejuvenate the blueberry plants. In practice, the commercial blueberry growers have found that two crops of berries between burns are all that the average land will produce profitably, and the three-year cycle usually keeps the weed bushes fairly well under control.

The life-cycle of *Rhagoletis pomonella* includes a firm obligate diapause in the pupal stage. In about 85 per cent of the population, diapause disappeared during the first winter, and most of the remainder emerged after the second winter. The pupal stage is spent in the soil, and most of the pupae occur in the top inch; nevertheless, the burning described above did not cause any deaths. The adults emerge from the soil and lay eggs inside the berries during June–July; the majority of the maggots leave the fruit and enter the soil to pupate during the first half of August.

The maggots of *Rhagoletis pomonella*, living inside the berries, may be sought out and oviposited in by certain predators, of which the braconid *Opius melleus* is the most important. The chance that a maggot will be found by a predator depends partly on the number of predators in the area and partly on the proportion of the blueberries infested by maggots. This is a situation quite analogous to the one we described in sections 11.12 and 10.321 in relation to *Anarhopus sydneyensis* and *Lygocerus* sp.

When the blueberry bushes are growing, as they did originally, as a sparse undergrowth among tall shrubs or along the margins of forests, they bear relatively few berries. In the absence of predators, the numbers of *Rhagoletis pomonella* would probably increase until nearly all the berries were infested. The population would still not be very dense, because there would not be many berries. But the numbers of the berries are not influenced in any way by the numbers of the maggots. However, the population is usually prevented from attaining this density by the activities of the predators.

The sequence of events in areas that are being "farmed" for blueberries is quite different. The account given by Lathrop and Nickels (1931, p. 263) is as follows:

> During the summer immediately following the burn there is a normal emergence of blue-berry flies on the burned-over areas, and the absence of a crop deprives the flies of berries in which to oviposit. This may be of little consequence on small patches, but where large areas are burned-over the flies must migrate considerable distances if they are to reach berries in which to oviposit. No maggots are produced on well burned land during the summer immediately following the burn, with the result that the population of the species is greatly reduced. The second summer following the burn is marked by the production of the first crop of berries, usually a heavy yield. The blueberry maggot has been "starved out" during the preceding season, and the beginning of the first crop year finds the population of the species at a low potential. There are two important sources of reinfestation of the new crop of berries: (1) migration of flies from unburned areas and (2) carryover of puparia in the soil from the second preceding season. Migration of flies is undoubtedly important on small burns of a few acres and near the margins of larger burned-over areas. Under usual conditions the migration of flies probably is not a very important factor on solid burns of 10 or 15 acres or more.

The population increases rapidly, because there is an abundance of food and the death-rate from predators is low indeed. This investigation did not reveal where the predators that were found in these populations came from. We do not know whether some of them also had spent two winters dormant in the soil or whether they had all flown in from outside. In either case their numbers were few and their chance of success slight. With such a large crop of berries and such a small proportion of them infested by maggots, the predators have little chance of finding many prey. The numbers of *Rhagoletis pomonella* continue to increase until the third summer after the "burn." Because the vegetation is usually burned again at this stage, the flies again become scarce.

Superimposed on this short-term rhythm in the numbers of *Rhagoletis pomonella*, which is determined directly by the frequency of burning, Lathrop and Nickels discovered a long-term trend which they described as follows:

If the land receives continued care, a period arrives, after several years, when the blueberry maggot population of the area reaches a maximum. At this point the land is relatively free from weed bushes, so an excellent stand of blueberries is supported, and berries are produced in abundance. However, there still remains a sufficient growth of "sprouts" and sweetfern to furnish protection to the adult flies. If the culture is carried beyond this stage, and the sprouts and sweetfern are completely removed from the land, there seems to be a tendency for the maggot population of the area to decrease, probably because of the excessive exposure of the adult flies to rain, wind and sunshine. Certainly, berries from blueberry land at its best development are seldom, if ever, found to have a high percentage of infestation.

If the blueberry land is neglected, the sprouts are not mowed, and the land is not burned-over for a period of years, the vegetation soon reverts to the tall shrub association. After a few years of neglect the yield of blueberries becomes insignificant. The percentage of berries infested by the maggots may be very great, but the total population of blueberry maggots on the area is small.

This study of the ecology of *Rhagoletis pomonella* in Maine shows us that where blueberries grow "naturally" as the sparse undergrowth among shrubs, the maggots are never numerous, because predators eat them. But in the culture of blueberries as practiced in Maine, although the amount of fruit is much greater, the number of maggots is small. This is not because predators eat the maggots but because of man-made catastrophes. In our way of looking at the ecology of an animal, this is an example of the numbers of an animal being determined by the number of suitable places for it, which, in turn, depends upon the activity of a nonpredator—man. The number of blueberries on an acre of land is not in any way determined by the density of the population of *R. pomonella;* nor is the frequency of the burning directly related to the number of flies. Neither can be properly described as a "density-dependent factor."

13.24   *The Causes of Deaths in a Natural Population of* Plutella maculipennis *near Pretoria*

The causes of deaths among a natural population of the tineid *Plutella maculipennis* were studied by Ullyett (1947). Samples were taken at weekly intervals during 1938–39; records were kept of the number of caterpillars on the plants and the causes of deaths occurring during the larval and pupal stages. Deaths among eggs and adults were not recorded, chiefly because of the technical difficulties. It was considered that these omissions did not seriously alter the final conclusions. Ullyett's (1947) paper, from which the following account was taken, dealt chiefly with the quantitative records made during 1938–39, but he stated that these results were confirmed by many less systematic observations made during the 6 years that the investigation lasted.

The life-cycle of *Plutella maculipennis* is not complicated by diapause. In the kindly climate of Pretoria, breeding continued throughout the year. A generation required about 14 days in midsummer and 21 days in midwinter. Generations overlapped, and all stages of the life-cycle were found on the plants at the same time. The females might, in favorable circumstances, lay up to 300 eggs, but in nature this potential was probably not very often attained. Ullyett considered the weather in some detail and concluded that the weather was not likely ever to be directly responsible for the deaths of many *P. maculipennis* in the region of Pretoria.

For food the larvae of *Plutella maculipennis* may eat the leaves of almost any crucifer. On the farm where Ullyett worked, cabbages and other crucifers were usually planted in long narrow plots. Cabbages were growing continuously on the same farm, though not on the same plot, throughout the year, except for 4 months during winter from July to October. However, there was no shortage of food for *P. maculipennis* at this time of the year, because there were plenty of *Nasturtium officinale, Brassica pachypoda, Lepidium* spp., and other crucifers, either cultivated in gardens or growing as weeds on this farm and elsewhere in its immediate vicinity. At no stage during the investigation did the population of *P. maculipennis* consume more than a small proportion of the food available to it in the area, so we can be reasonably certain that its rate of increase was never seriously reduced by an absolute shortage of food. But the periodic harvesting of each crop of cabbages, the planting of new batches on different plots, and the complete absence of cruciferous vegetables from all the plots during 4 months meant that the distribution of the food was constantly changing. The distances involved may have been small relative to the powers of dispersal of *P. maculipennis*, for Ullyett did not consider this as a serious hindrance to its multiplication (sec. 11.12).

In a kindly climate and with plenty of food and places to live, it was to be expected that the numbers of *Plutella maculipennis* would be determined largely by the other organisms associated with them. Ullyett reared 18 species of Hymenoptera and one species of Diptera from the larvae and pupae of *P. maculipennis*. Nine of them laid their eggs in or on the larva, and four oviposited in or on the pupa; all of them lived inside the body of the prey, that is, they were, in the language of economic entomology, "internal parasites" of *P. maculipennis*. The other six came into our category *e* of section 10.01; they were predators of predators in the environment of *P. maculipennis*. Rather little was known about most of the species in this list. Some were found only rarely, and it is likely that they were casual predators of *P. maculipennis*, preferring some other unknown species. But two species, *Angitia* sp. and *Apanteles halfordi*, were known to prey chiefly, if not exclusively, on *P. maculipennis*. The former was the only one that ever became abundant; in Figure 13.14 the numbers represented by curve *C* were made up chiefly by this species.

The pupae of *Plutella maculipennis*, especially those that were fastened to the outer leaves which touched the ground, were often eaten by beetles of the family Staphylinidae. The staphylinids lived in the soil, and the greater part of their food consisted of a variety of soil-inhabiting insects; hence they were not dependent on *P. maculipennis*. It so happened that they occasionally became quite numerous in the soil under the cabbages, and they sometimes were responsible for a large proportion of the weekly deaths of *P. maculipennis*. The cabbages sometimes harbored large populations of aphids, which attracted

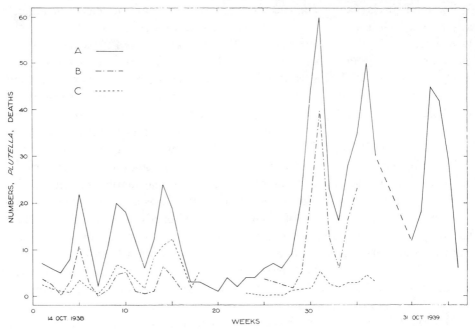

Fig. 13.14.—*A*, the numbers of *Plutella maculipennis* in a garden near Pretoria, based on weekly counts of the numbers of larvae and pupae on a sample of cabbage leaves; *B*, the weekly deaths due to facultative predators; *C*, the weekly deaths due to obligate predators. (After Ullyett, 1947.)

their own group of predators. Larvae of certain species of hover flies (Syrphidae) and lacewings (Hemerobiidae or Chrysopidae) and the nymphs of an anthocorid bug were the most important. It so happened that these predators would sometimes consume all or most of the stock of their preferred food (aphids) on a cabbage before they were themselves fully grown. In these circumstances they would turn to the larvae of *P. maculipennis*, and on occasion a high proportion of the weekly deaths was attributed to them. Other, less important, facultative predators include spiders, wasps (*Polistes*), and birds.

These facultative predators had certain qualities in common. Since *Plutella maculipennis* did not constitute a staple part of their diet, their numbers fluctuated independently of those of *P. maculipennis*. Since the caterpillars and pupae of the cabbage moth were not the preferred food of this group of predators, they were not likely to be eaten in numbers unless they were numerous and easy to find. An exception to this occurred when the predators of aphids found themselves deprived of their favorite food. Being wingless, they were unable to leave the cabbage in search of other food, so they ate what they could find. For these reasons the deaths attributed to the activities of this group of facultative predators fluctuated widely and sporadically (Fig. 13.14, curve *B*).

Twice during the period shown in Figure 13.14, widespread outbreaks of a fatal disease caused by the fungus *Entomophthora sphaerosperma* swept through the population of *Plutella maculipennis*, leaving only two or three larvae alive on each cabbage plant. During the first outbreak the weekly counts were maintained, and in Figure 13.14, curve *A* shows that the population remained at this low level for nearly 10 weeks. Precise numbers were not recorded during the second outbreak, but Ullyett (1947, p. 99) wrote: "After the initial peak of the larval infestation had been passed, a period of heavy, continuous rains supervened which prevented field work. This was immediately followed by another epidemic of the fungus disease which reduced the population to a low level comparable with that reached in Period II."

Between outbreaks very few individuals died from this or any other sort of disease. But the fungus must have been present and widespread, because outbreaks of the fungal disease invariably occurred whenever there had been enough rain to keep the surfaces of the leaves of the plants continuously wet for 3 or 4 days running. Light continuous rain was more effective than heavy intermittent rain; even continuous heavy mist would do. This disease was observed to break out with equal severity in a number of different populations of different densities, and Ullyett concluded that the onset of the disease was independent of the density of the population at the beginning but depended only on the weather. While the weather remained favorable to the fungus, the numbers of *Plutella maculipennis* remained at a consistently low level. Ullyett wrote: "The extent to which the population is reduced has been approximately the same on all occasions when the duration of the disease has been comparable. There are indications that this agreement would be even more remarkable if the density of the host population could be correlated with the total surface area of the plant." Once the rain abated and the surface of the plants remained dry for considerable periods between showers, the panzoötic came to an end; and the numbers of *P. maculipennis* increased at a great rate. Figure 13.14, curve *A*, shows the increase quantitatively for the period after the first outbreak of disease, but the curve relating to the period after the second outbreak is qualitative only. Precise records were not kept, but Ullyett commented: "A repetition of the previous events occurred and the end of the following year's observations was marked by an exceedingly heavy infestation of all cruciferous crops which were then on the ground. In this case the resultant damage was so severe that cabbage and cauliflower crops were unmarketable and were fed to stock."

The rate at which the population increased during the intervals between outbreaks of disease depended partly on the fecundity of the moths and the speed of development of the immature stages and partly on the activities of predators. The deaths caused by facultative predators (curve *B* of Fig. 13.14) doubtless retarded the rate of increase somewhat, but they did not prevent the population from becoming so dense that cabbages and cauliflowers were unmarketable and were fed to stock. Since *Plutella maculipennis* were not the favored food of any of these groups of facultative predators and since the numbers of the predators were largely independent of those of *P. maculipennis*, there are good grounds for expecting that these predators would not ordinarily keep *P. maculipennis* scarce. On the other hand, the obligate predator, *Angitia* sp., might be expected to do this, provided that its rate of increase and capacity for increase were adequate (sec. 10.321).

The same disadvantage associated with changing distribution of crucifers on the farm which we mentioned above in relation to *Plutella maculipennis* applied also to *Angitia* sp., which had to find their prey as they spread out to colonize the plants in each new situation. The evidence which we reviewed in section 10.321 would suggest that a small discrepancy in the dispersive powers of prey and predator might make a big difference to the numbers which the prey can maintain in a particular area. Ullyett did not mention this point explicitly, but he implied from his discussion of the results that *Angitia* sp. and the other species represented in curve *C* of Figure 13.14 possessed the necessary qualities for keeping *P. maculipennis* scarce.

The predators of this group were probably chiefly responsible for the low numbers of *Plutella maculipennis* recorded during the first 15 weeks of the period shown in Figure 13.14. But they clearly did little to check the multiplication of their prey during the periods which followed the disappearance of the epizoötic of fungal disease. While the disease was widespread, the prey suffered severely, but the predator suffered even more severe losses: not only were many killed by the disease, but those that survived suffered from a shortage of food. The predators started from a relatively low level after the epizoötic abated. As a consequence, their rate of increase was at first very slow relative to that of the prey; they multiplied steadily but did not become numerous enough to effect any appreciable reduction in their supply of food before catastrophe, in the form of another outbreak of disease, overtook them once again.

Several conclusions of general interest emerged from this investigation: (*a*) At no time did *P. maculipennis* become numerous enough to eat more than a small proportion of the stocks of food available to it. (*b*) The caterpillars were reduced to minimal numbers during epizoötics of fungal disease, the average number of caterpillars per plant being usually 2 or 3 while the epizoötic lasted; but we do not know how this particular density was determined. (*c*) Outbreaks of disease were occasional and temporary; their frequency and duration were determined entirely by the weather. (*d*) The maximal numbers of caterpillars were attained shortly after the epizoötic had abated. (*e*) The numbers prevailing at other times, that is, when there was a prolonged interval between outbreaks of disease, probably depended largely on the activities of predators, especially of the obligate predator *Angitia* sp.

From what has been said, it is clear that in the Pretoria district the relative importance of diseases and predators in determining the prevailing numbers of *Plutella maculipennis* depends on the frequency and duration of spells of wet weather. This matter was not investigated by Ullyett. For the short period covered by Figure 13.14, disease was undoubtedly the more important in relation to both maximal and minimal numbers.

13.3 SOME EXAMPLES WHICH ILLUSTRATE THE IMPORTANCE OF STUDYING THE PLACES WHERE THE ANIMALS MAY LIVE

The first example in this section may seem unduly simple. So far as we can tell, the distribution and abundance of the two species of nematodes *Trichostrongylus vitrinus* and *T. colubriformis* is determined entirely by what we called in sections 12.0 and 12.1 the "qualitative aspect" of the place where an animal may live. By way of contrast, the ecology of *Ixodes ricinus*, which forms the second example, is more complex than any other example which we have discussed. It includes a nice interaction between moisture and "place in which to live"; food is also important in a subtle way, and so is the number of other animals of the same kind. In the third example there is a nice interaction between "food" and "place in which to live"; moisture is also important in its own right. Finally, we discuss the ecology of the bobwhite quail; this is interesting because it is part of the original work on which Errington based his concept of "carrying capacity," which we have had occasion to mention in sections 10.322 and 12.31.

13.31 *The Distribution and Abundance of the Nematodes,* Trichostrongylus vitrinus *and* T. colubriformis *in the Gut of the Sheep*

Several species of nematode worms commonly occur on the surface of the mucous membrane of the gut of the sheep. More often than not, several species occur in the one animal. Some of them are found only in the abomasum, others only in the small intestine. The distribution is even more specific than this, as certain species tend to concentrate in specific sections of the small intestine (Tetley, 1937; see also Fig. 13.15). The three species shown in Figure 13.15 reach their maximal numbers in the first few feet of the small intestine, becoming fewer toward the pylorus and in the lower reaches of the duodenum (Somerville, personal communication). Food is undoubtedly present in abundance, and these striking distributions must reflect a gradient in some quality of the gut; it is almost certainly a chemical gradient, perhaps a gradient of pH.

The information summarized in Tables 13.08 and 13.09 reveal two interesting

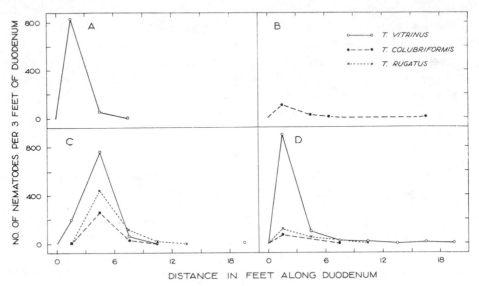

FIG. 13.15.—The distribution and abundance of three species of *Trichostrongylus* along the length of the duodenum of a sheep. Each graph shows the distribution in a single sheep. *A*, *T. vitrinus* by itself; *B*, *T. colubriformis* by itself; *C* and *D*, *T. vitrinus*, *T. colubriformis*, and *T. rugatus* together. (From Somerville, personal communication.)

features about the distributions of *Trichostrongylus vitrinus* and *T. colubriformis:* (*a*) the two species occur in the same part of the gut, and (*b*) the presence of other species in the gut seems not to modify in any way either the

TABLE 13.08*
DISTRIBUTION AND ABUNDANCE OF *Trichostrongylus vitrinus* IN GUT OF 10 SHEEP

OTHER SPECIES PRESENT		DISTRIBUTION IN DUODENUM IN FEET (BEGINNING OF DUODENUM = 0)		MAXIMAL NO. PER 3 FEET OF DUODENUM
No.	Species	Major Distribution	Subsidiary Distribution	
0..........	....	0–6	....	820
1..........	*T. colubriformis*	{0–6 0–12	....	270 360
2..........	{*T. colubriformis* *T. rugatus*	{0–9 0–9 0–12 0–6 0–9	18–31 15–18 12–21, 30–36 36–42	760 900 150 60 440
2..........	{*T. rugatus* *T. probolurus*	0–6	....	30
3..........	{*T. colubriformis* *T. rugatus* *T. probolurus*	0–9	15–18	40

* From Somerville, personal communication.

TABLE 13.09*

DISTRIBUTION AND ABUNDANCE OF *Trichostrongylus colubriformis* IN GUT OF 10 SHEEP

OTHER SPECIES PRESENT		DISTRIBUTION IN DUODENUM IN FEET (BEGINNING OF DUODENUM = 0)		MAXIMAL NO. PER 3 FEET OF DUODENUM
No	Species	Major Distribution	Subsidiary Distribution	
0..........	....	0–6	15–18	100
1..........	*T. vitrinus*	{0–6	.....	70
		{0–12	.....	60
2..........	{*T. vitrinus*	{0–6	9–15	75
	{*T. rugatus*	{0–9	.....	100
		{0–9	.....	260
		{0–6	.....	90
		{0–6	.....	30
3..........	{*T. rugatus* }	{0–12	12–21	3,000
	{*T. probolurus* }	{0–18	21–30	2,250
	{*T. vitrinus* }			

* From Somerville, personal communication.

distribution or the abundance of either of these two species. If one species interfered with another, it might be expected that the distribution of each species would become restricted in the presence of others (sec. 10.213; and Fig. 10.09). This is clearly not the case.

This example brings out two interesting points: (*a*) a gradient in some attribute of the area where a population is living may be reflected in the abundance of the animals, and (*b*) at least in the special circumstances represented by the gut of the sheep, a number of closely related nonpredatory species can live crowded together in a restricted space.

13.32 *The Distribution and Abundance of* Ixodes ricinus *in Pastures in Northern England*

The tick *Ixodes ricinus* is regarded as a serious parasite of sheep in Britain, especially as it is known to be a vector of disease. It is found on pastures of the uplands but not of the lowlands, where it might have been expected to occur if one considered only climate and the presence of its chief host the sheep. On the upland pastures it is distributed in an irregular way which was for a long time rather puzzling. The ecology of *I. ricinus* was thoroughly investigated by Milne (1943, 1944, 1945*a*, *b*, *c*, 1946, 1947*a*, *b*, 1948, 1949, 1950*a*, *b*, 1951, 1952). There are also several other papers, mention of which may be found in those we have cited. The distribution and abundance of *I. ricinus* are no longer puzzling, and the explanation includes more especially a consideration of (*a*) the ticks' requirements for water and how these may be satisfied in certain sorts of places but not in others and (*b*) the hazards associated with the ticks' "search" for food and mates and how these may be increased by a scarcity of either sheep or ticks.

The life-cycle of *Ixodes ricinus* was described in section 12.22; it occupies 3 years. Except for a brief period of activity during the spring,[2] the tick spends the whole of each year lying concealed at the base of the vegetation near the surface of the soil. One huge meal of blood is taken during the spring, but for the rest of the time the tick neither feeds nor drinks. The unfed larva or nymph loses water by evaporation when it is exposed to air in which the relative humidity is less than 88 per cent; the unfed adult loses water to air in which the relative humidity is less than 92 per cent; but ticks in either stage may absorb water through the cuticle from air in which the relative humidity exceeds these values. So a tick has little chance of remaining alive during the periods which elapse between meals unless it happens to be in a place where the air remains nearly saturated with water vapor continuously throughout the year. The physiology of the tick in relation to moisture is discussed in sections 7.213 and 12.22.

Since the tick can crawl only very feebly, it invariably burrows into the vegetation within an inch or two of where it drops from the host. So the place where the tick spends the interval between meals depends not at all on its own activities (except that it may crawl up or down the grass stems, coming closer to, or going farther away from, the surface of the soil) but is determined entirely by the movements of the host. If a tick is to survive to contribute progeny to the next generation, it must meet with good fortune, with regard to the place where it drops from the host, not once but three times during its life. A tick which falls from a sheep will die from desiccation unless it is lucky enough to fall where there is a dense mat of vegetation and dead organic matter on top of the soil. If it is lucky enough to fall into such a situation, it will burrow down into the thick central part of the mat where the air is most moist. It remains there until it is ready for the next meal of blood (sec. 12.22).

The pastures which grow on uplands, especially on the lower slopes of hills, include coarse grasses and bracken, which form a dense, tangled mat of vegetation and dead organic matter next to the soil. This mat is often several inches thick, and it is likely to remain continuously moist throughout the year. An engorged tick, dropping from its host into this mat, is not likely to die from desiccation before the next meal is due. On the other hand, pastures on the lowlands usually consist of finer grasses, more closely grazed. Except perhaps for a few local situations, they contain no place where the air is likely to remain saturated with water vapor, or nearly so, continuously throughout the year. The chance that a tick dropping from a host into one of these pastures would find a safe place to live is remote; the chance that it would do so three times during its life is so remote that none ever survives in these lowland pastures, despite the fact that sheep infested with ticks are often brought to

2. In some districts the ticks are active during autumn as well as spring; the following account refers especially to those districts where they are active only in the spring.

them. Breeding populations of *Ixodes ricinus* persist only in the rough pastures of the uplands.

Milne sampled the populations of ticks by dragging a blanket over plots of about 100 square yards. It was quite impracticable to look for the ticks while they remained inactive at the base of the vegetation. So the sampling had to be done during the period between April and June, when the ticks were to be found on the tips of the grass stems waiting to attach themselves to a passing host. Although ticks were to be found on the grass stems at any time during this period, the active stage for any individual did not exceed about 12 days. So the numbers on the grass stems and hence the numbers trapped on the blanket depended partly on the numbers in the area and partly on their activity. As in all methods of simple trapping, so it was in this case: there was no easy way of discriminating between these two components. But Milne equalized the component for activity by taking simultaneous samples on the plots that he wished to compare and by spacing the samples uniformly through-out the period when the ticks were active. Usually, because of the labor in-volved, a limited number of samples was taken from each plot, and these were spaced rather widely in time, so that they were likely to contain an unspecified, but approximately constant, proportion of the total population on the plot. That is to say, his comparisons were based on relative, not absolute, numbers. This is, of course, a characteristic of all methods based on simple trapping.

Milne compared the numbers of ticks at different altitudes on one farm, where they had been established for many years and where the population was relatively dense. The chief difference in the pastures at different altitudes was in the mat, which was thicker and denser on the lower slopes and shallower near the tops of the hills. The results summarized in Table 13.10 show that the numbers were generally fewer on the higher slopes and virtually absent

TABLE 13.10*
NUMBER OF NYMPHS OF *Ixodes ricinus* AT DIFFERENT ALTITUDES IN SAME "GRAZING"

Location	Date	Altitude (Feet)	Mean Nymphs per "Drag"
Hill A...........	April 2	500	1.5
		900–1,100	0.0
	April 12	500	9.0
		800	2.0
Hill B...........	April 25	400–500	19.0
		500–600	19.0
		600–750	0.0
Hill C...........	June 9	450–500	5.7
		600–800	1.0
		900–1,050	0.3

* Plots were chosen in each case so that the pasture would be typical of the altitude. Data from Milne (1946).

from the tops of the hills. Another experiment showed that this was due to the thinning-out of the mat on the higher slopes.

In some situations, because of minor variations in topography and soil, there were small areas at high altitudes where the mat was thick and dense. In these places the ticks were just as numerous as they were in the thick mat at the lower altitudes (Table 13.11).

TABLE 13.11*

NUMBER OF NYMPHS OF *Ixodes ricinus* AT DIFFERENT ALTITUDES ON
SAME "GRAZING" AS IN TABLE 13.10

Date	Height	Vegetation	Mean Nymphs per "Drag"
May 2	⎰450	Thick	1.80
	⎱750	Thick	3.67
	650–900	Thin	1.00
	1,000–1,050	Thin	0.50
May 9	800–1,000	Medium	1.00
	1,100	Medium	2.00
	500–600	Thick	4.50
	1,050	Thick	14.00

* In this case the plots at the higher altitudes were chosen especially from local situations where the mat was unusually thick. Data from Milne (1946).

In another experiment the number of ticks in a "sheep lair" were compared with the number in a representative plot of the pasture where the sheep grazed during the day. The lair was a small area near the top of a hill, where the sheep always went at night to sleep. The pasture on the lair was sparse and the mat thin. Since the sheep spent a greater proportion of their time on the lair than on any other area of comparable size in the meadow, it might have been expected that more engorged ticks would drop from the sheep on the lair than on the pasture elsewhere. Milne found that his samples yielded 0.21 nymphs per "drag" of the blanket on the lair and 5.18 nymphs per "drag" on a representative plot of pasture on the lower hillside. He concluded that the ticks were scarce on the lair because few could survive in the absence of an adequate mat. He added: "Possibly replete ticks do drop off in large numbers on the lairs but most should be unable to breed. Hence the lairing habit of hills sheep probably causes the deaths of many ticks."

Even a heavily infested sheep never carried enough ticks to suggest that food was ever scarce in the absolute sense. Nevertheless, shortage of food in the relative sense in which we discussed it in section 11.12 was probably the chief cause of deaths among ticks situated in good rough pasture where there were plenty of suitable places for them to live. For example, Milne estimated that the population on half of a certain farm included between 70,000 and 116,000 females of *Ixodes ricinus;* there were rather fewer on the other half of the farm, but these were not estimated. During the period when the ticks were actively seeking hosts, this farm was stocked at the rate of about one sheep to the acre. This is the highest rate of stocking likely to be found on upland farms. During this season the entire flock picked up between 15,000 and 45,000 females. With the most conservative estimate possible from these figures, over 60 per cent of the adult ticks on this farm died from lack of food because a tick cannot live through a second summer without food. If the deaths among the larvae and nymphs occurred at the same rate, then the total deaths from this cause during a full generation would be of the order of 94 per cent. It would, of course, be higher on a farm that was more lightly stocked.

If a tick is to become attached to a host, it must first be sitting on the tip of a grass stem at the time when the host brushes past that stem. With few sheep per acre, the frequency with which any one grass stem is brushed by a sheep may be small. Nevertheless, a tick might have a good chance of being picked up if it were able to remain a long time on the tip of a grass stem. But this is just what it cannot do. While a tick is sitting on the tip of a grass stem, it is losing water by evaporation to the air. It is likely to die from desiccation if it remains there for more than 5 days. Before this period has elapsed, if it has not been picked up by a host, it usually returns to the mat near the soil, where it recuperates. On the average, a female tick spends about 9 days exposed on the tip of a grass stem. Usually this is made up of several separate periods of 2 or 3 days each. Including the periods spent in the mat between visits to the tip of a grass stem, the period of activity of any one tick usually extends over no more than 4 weeks (Lees and Milne, 1951). Those which fail to attach themselves to a host during this period die.

In most places ticks become active for a period during autumn and again during spring; in some districts activity is restricted to the spring. Figure 13.16

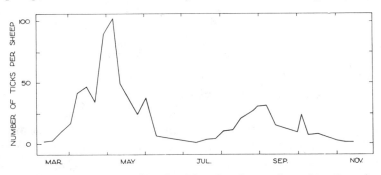

FIG. 13.16.—The average number of *Ixodes ricinus* found on a sheep throughout the year on a farm (Crag) in Cumberland. (After Milne, 1945a.)

shows the average number of females found attached to sheep at different dates during 1940 and 1941 in a pasture at Crag. Milne withheld sheep from an experimental plot until 3 weeks after the time for the usual peak of activity in the spring. By the end of the season, the sheep on this plot had picked up only half as many ticks as those on the control plots, showing that about half the population must have ceased activity before the sheep were returned to the pasture. Activity ceased on both the experimental and the control plots at about the same date. This experiment showed that the dates when the ticks become active and cease being active are independent of the presence of the sheep. They are doubtless determined by the progression of the seasons (secs. 8.1, 8.12).

Sheep are not the only hosts of *Ixodes ricinus;* almost any mammal or bird which frequents pastures can probably serve as food for the ticks. Milne recorded as hosts 29 species of mammals, 39 birds, and 1 reptile. The importance of one species of host relative to another depends both on the readiness with which ticks are collected by an individual and on the number of individuals in the area.

In comparing the readiness with which different species of hosts collected ticks, it was necessary to count ticks on different species taken from the same place at the same time, because the number of ticks on the ground varied from place to place and the number of ticks which were active and ready to cling to a passing host varied with the seasons. Milne expressed the results of such counts for a number of species of hosts in terms of "sheep equivalents." The hedgehog, for example, had a "sheep equivalent" of 0.1353. This means that one hedgehog carried as many ticks as 0.1353 of a sheep. The "sheep equivalents" of a number of hosts are shown in Table 13.12.

TABLE 13.12*
"SHEEP EQUIVALENTS" OF NUMBER OF ADULT HOSTS OF FEMALE TICKS

Roe deer	0.1396	Otter	0.0324
Hedgehog	.1353	Fox	.0233
Brown hare	.0771	Pheasant	.0202
Stoat	.0547	Red grouse	.0017
Badger	.0544	Magpie	0.0015
Rabbit	0.0014		

* Mean for five farms in the north of England. After Milne (1949).

Even the "best" wild animal is far behind the sheep as a host. In general, a host which covers more ground will pick up more ticks. The smaller the animal, the less the area of body which is in contact with the vegetation and generally the less the distance traveled in search of food. This means that small animals usually "sweep" a smaller area of ground than large ones do. There are, of course, exceptions, since habits of species vary and some animals touch the vegetation more intimately than others. For example, the mole is larger than the shrew, but it usually harbors fewer ticks than the shrew because the mole spends more time underground and so has less opportunity of meeting ticks.

The figures in Table 13.12 do not indicate the relative importance of these different animals as hosts, for that is also dependent upon the abundance of the hosts. Milne estimated the relative abundance of these various hosts on five farms in England. The mean numbers for these five farms, which had an average acreage of 1,500 acres, is shown in Table 13.13. Some of the animals shown in Table 13.13 spent very little time on the pasture. The otter, for

TABLE 13.13*
ESTIMATED NUMBERS OF HOSTS OF FEMALE TICKS ON 5 FARMS IN NORTH
OF ENGLAND WITH AVERAGE ACREAGE OF 1,500 ACRES

Rabbit	692	Hedgehog	12
Sheep	590	Fox	4
Red grouse	322	Roe deer	3
Pheasant	42	Magpie	3
Cattle	30	Badger	2
Brown hare	20	Otter	2
Stoat	16		

* The figures give the mean number of hosts per farm. After Milne (1949).

example, spent most of its time in the water; it visited the farms at infrequent intervals. The badger likewise must be quite unimportant as a host because of its rareness. A pair of badgers in hill and moorland country are said to need at least 500 acres for their upkeep.

The importance of the different species of hosts relative to the sheep can be obtained by multiplying the "sheep equivalent" of each species (Table 13.12) by the number of individuals (Table 13.13). The product is a measure of the relative number of ticks carried by the different hosts and is called by Milne the "population sheep equivalent" of the wild hosts (Table 13.14). The

TABLE 13.14*
"POPULATION SHEEP EQUIVALENTS" OF WILD HOSTS, INDICATING RELA-
TIVE NUMBERS OF FEMALE TICKS CARRIED BY WILD HOSTS

Hedgehog	1.62	Roe deer	0.42
Brown hare	1.54	Badger	.11
Rabbit	0.97	Fox	.09
Stoat	0.87	Otter	.06
Pheasant	0.85	Magpie	0.01
Red grouse	0.55		
		Total	7.09

* After Milne (1949).

total wild hosts on the five farms to which Table 13.14 referred represented an average of 7.09 sheep per farm. This was 1.19 per cent of the total sheep population per farm. In other words, on these farms the sheep provided the food for nearly 99 per cent of the adult females that found hosts of any sort. Corresponding figures for nymphs were worked out on another farm. Sixteen species of wild hosts on this farm were estimated to have a total "population sheep equivalent" of 30.71 sheep. There were 590 sheep on the farm, and they provided the food for 95 per cent of the nymphs that found hosts of any sort. These experiments showed that the ticks were dependent almost entirely on the sheep for food and a place on which to mate. The presence of wild hosts did little to ameliorate the tick's chance of finding food and a mate.

We discussed in section 9.13 how the birth-rate in populations of *Ixodes ricinus* may be reduced by the failure of the ticks to find mates. Since the sexes can meet only on a host, the chance that a female will be fertilized depends on the number of ticks likely to be picked up by one host. This depends largely on the density of the population of ticks; increasing the number of hosts on a particular area may increase the tick's chance of finding food, but not its chance of finding a mate. This is at the root of the explanation of the erstwhile puzzling observation that many farms on which the pastures are undoubtedly well suited to *I. ricinus* have remained for many years free from ticks, despite the fact that ticks must have been repeatedly brought in by wild hosts. But so few are likely to be introduced at any one time in this way that their chance of colonizing the new area must be exceedingly small. The presence of sheep in their usual numbers on these pastures showed that the ticks' failure to become established was due not to the hazards associated with finding food but to the very high risk of not finding a mate. Shortage of mates is associated with a too sparse population of ticks; shortage of food is not associated with a too dense one; the ticks in the same population suffer at one and the same time from the difficulties of finding food and the difficulty of finding a mate. This may seem a paradox to those who always associate shortage of food with "intraspecific competition," but food is not a "density-dependent factor" when it is scarce in the relative sense (sec. 11.12). For another example of this see *Glossina morsitans* (sec. 13.33).

Other hazards which might reduce *r* in natural populations of *Ixodes ricinus* include predators, which are not very prominent, and certain risks which the tick encounters while it is on the host. It may "get lost" in the fleece of the sheep; some parts of the fleece are lethal to the tick; if it happens to become attached in a lethal region, it may die there. The tick has the best chance of surviving on the bare and hairy parts of the body. The tick may be killed by its host after it has become attached. Birds and the smaller mammals "detick" themselves; with birds especially, ticks are found only on the upper regions of the neck and head, where they cannot be reached by the beak; the ones which attach themselves elsewhere on the body are invariably eaten. Larger animals, irritated by the ticks, rub themselves against walls and posts; and some ticks are killed in this way. The skin of a sheep which has been repeatedly infested by ticks may acquire a reaction which may prevent some ticks from engorging or may even kill them. But these risks are less important than the others which we have mentioned.

This study of *Ixodes ricinus* is one of the nicest pieces of ecology that we know. The place where a tick may be living is important because of an interaction with one component of weather, namely, moisture. Food is important: the tick is so lacking in dispersive ability that it suffers severely from shortage of food, notwithstanding the presence of plenty of food in the area if only the tick could find it. The tick also suffers from too few other animals of its own kind in its environment; this is also largely due to its poor dispersive ability. It is noteworthy that Milne was able to offer a thorough and adequate explanation of the distribution and abundance of *I. ricinus* without referring to the dogma of "density-dependent factors."

13.33 *Some Points from the Ecology of Four Species of Tsetse Flies*, Glossina *Spp.*

"Tsetse fly" is the name given to a group of bloodsucking flies of the genus *Glossina* which are found in forests, woodlands, and other shady places in Africa. They have been extensively studied because they carry serious diseases among men and domestic animals. A number of papers dealing with the ecology of *G. morsitans*, *G. tachinoides*, *G. palpalis*, and *G. longipalpis* have been published by Jackson (1930, 1933, 1936, 1939, 1949), Nash (1930, 1933a, b, 1937, 1948), Gaschen (1945), Potts (1937), and Zumpt (1940).

One interesting aspect of this work is that the absolute numbers of flies in certain areas have been estimated by the sampling method known sometimes as "the Lincoln Index" but better called "the method of capture, marking, release, and recapture." The principle of the method is simple. A sample is collected from the population to be measured. The individuals are marked so that they may be recognized again and are then released in such a way that they may be expected to distribute themselves at random with respect to the rest of the (unmarked) individuals in the population. At a later date another sample is taken at random from the population, and the numbers of marked and unmarked individuals in it are recorded. It is important that either the initial marking or the subsequent catching be done evenly over the area selected for study. The second sample will contain some marked and some unmarked individuals. Provided that (*a*) the marked individuals had redistributed themselves at random to the unmarked ones; (*b*) the marked ones are neither more nor less readily caught than the unmarked ones; and (*c*) between the times of release and recapture there have been no gains or losses by births, deaths, or migration, the total population in the area equals

$$\frac{\text{Number marked at first} \times \text{total marked and unmarked recaptured}}{\text{Number of marked ones recaptured}}.$$

In nature, condition *c* is never likely to be fulfilled, but it is the great merit of this method that, by suitably designing the experiment and making the appropriate extensions to the fundamental equation, one can measure not only the total population in the area but also the rate at which it is changing. Moreover, the rate of change may be divided into its four components of births, deaths, immigration, and emigration. A full account of the method as it was used for *Glossina morsitans* may be found in three papers by Jackson (1933, 1936, 1939). Dowdeswell, Fisher, and Ford (1940) discussed the method in some detail in relation to their study of the numbers of a natural population of the butterfly *Polyommatus icarus*. A general account of the method was also given by Ford (1945a, p. 270).

Evans (1949) attempted to use the method with a natural population of house mice, *Mus musculus;* but he had to abandon it, because he found that there were some individuals which liked to be trapped and others which were especially shy of the traps. The former had a greater chance of being marked in the first instance and of being recaptured. There was no way of getting over this difficulty. With other species, other sorts of difficulties may arise which make it doubtful whether conditions *a* and *b* have been fulfilled. So the method is not, unfortunately, available for general use in the study of natural populations.

In what follows we shall mention a few interesting points about several species of tsetse flies without going deeply into the ecology of any one.

There is no diapause or quiescence in the life-cycle of *Glossina morsitans* in West Africa, and new individuals are being added to the population throughout the year. For brief periods at certain seasons, births exceed deaths, and *r* is positive, resulting in a temporary increase in numbers; for the remainder of the year deaths exceed births; *r* is negative, and the numbers are declining.

The larva of *Glossina morsitans* does not lead an independent existence: it is nourished to maturity inside the body of the female. The fully mature larva is placed by the female on the surface of the soil or among litter, especially under bushes or against logs. The larva crawls into the soil and pupates. The pupal stage may last for several weeks. The pupa may die if the place where it happens to be should become too dry or too wet. The northern limit to the distribution of *G. morsitans* is determined largely by the survival-rate among pupae which have to develop during the dry season; the limit corresponds rather closely to the 30-inch isohyet (Fig. 13.17), which is associated with a dry season of 6 or 7 months.

In the northern parts of their distribution *Glossina morsitans* may increase in numbers during the wet season as well as during the first half of the dry season. But as the dry season advances, *r* becomes negative. The flics begin to evacuate the drier parts of their range in the open savannah woodland early in the dry season, slowly becoming restricted to the islands of forest or the denser vegetation along the rivers. By March, which is the hottest and driest month in Nigeria, the flies are to be found only in these moister, more shady areas. At this time of the year the flies enter adjacent woodlands to hunt only in the mornings and evenings. As soon as the rains begin again, *G. morsitans* begin to disperse into the savannah woodland, individuals traveling as far as 8 miles from the places where they had spent the dry season. During the dry summer, larvae are deposited only in thickets, but during the wet season they may be placed under logs in open savannah. New generations

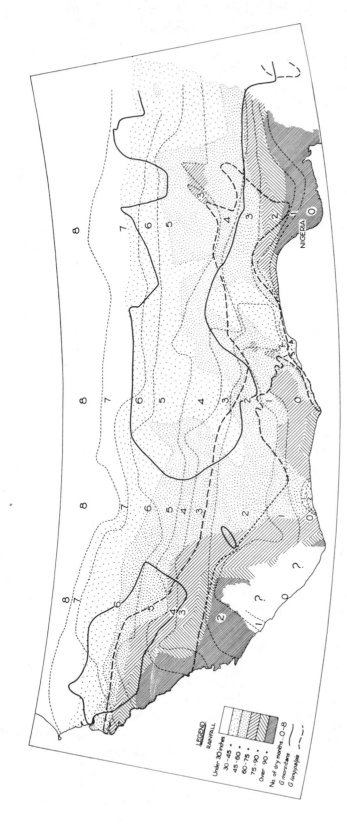

Fig. 13.17.—The distributions of *Glossina morsitans* and *G. longipalpis* in West Africa in relation to annual rainfall and the number of "dry" (less than 1 inch) months in a year. (After Nash, 1948.)

breed in the newly invaded areas, and, in due course, the flies disperse far and wide through the savannah woodland. A period is set to this multiplication and expansion by the return of the dry season.

The southern limit to the distribution of *Glossina morsitans* seems not to correspond to any isohyet. There is evidence that excessive wetness may be harmful to the pupae; in the southern (wetter) parts of their distribution deaths exceed births during the wet season, and r remains negative; the first part of the dry season provides the only period when r remains positive. Furthermore, in East Africa, in Tanganyika, the distribution of the species may expand during a run of dry years into vegetational zones which would normally not be inhabited by them. Nevertheless, the limits of their distribution on the wet side are probably determined less by the direct influence of moisture on the insect itself than by the influence of moisture on the distribution of certain types of vegetation.

In Nigeria the vegetation runs in bands that are roughly parallel to the coast. The coast is fringed by mangrove swamps and swamp forests. The sequence that is found as one proceeds inland is evergreen forest, mixed deciduous forests, heavy savannah woodland with islands of mixed deciduous forest, and, finally, far inland, light savannah woodland. Narrow belts of vegetation typical of wetter zones follow the rivers into the drier areas. Figure 13.17 shows that *Glossina morsitans* is absent from the heavy evergreen forests of the south; the precise explanation for this seems not to have been worked out yet. The influence of vegetation is better understood for some other species, especially for *G. tachinoides* (see below).

The adults of *Glossina morsitans* feed only on the blood of ungulates, especially antelopes, and their distribution may impose a further limit on the distribution of the fly. The numbers of *G. morsitans* are not likely ever to be so great that it suffers from a shortage of food in the absolute sense, but food may sometimes be difficult to find when antelopes are scarce. The following is the gist of a hypothetical example which Jackson (1937, p. 886) made. Suppose that there are 100 tsetse flies in an area which also supports just enough antelopes to enable the flies to maintain their numbers indefinitely without increase or decrease. This does not mean that the total weight of blood is just adequate for the needs of 100 flies but rather that the antelopes are just numerous enough to insure that the average tsetse fly meets with food often enough for it to produce offspring at the rate required to match the death-rate in the population. Suppose that 900 new flies are introduced to the area. The fecundity of each newly arrived fly, depending as it does on the frequency with which the fly meets an antelope, will be just the same as that of the original inhabitants, and the population will continue to maintain itself at a steady level. This is because, even with the larger number of flies, there is still no shortage of food in the absolute sense. Jackson expressed the same point in different words: "It [food] may no more increase its action on a rising population of the tsetse fly than does climate, nor temper its severity towards a diminishing community. [In other words, food is not operating as a 'density-dependent factor.'] It seems that there is no pressure of numbers in the ordinary sense because there is probably no competition for food, and certainly none for shelter, as the writer's experiments indicate that the flies are so sparse that there can be no physical crowding even when the apparent density is comparatively very high." Elsewhere in the same paper Jackson pointed out that there is no evidence that predators or disease influences numbers to any extent, and he concluded: "There remains the possibility that (at moderate densities at least) there are no dependent factors at all acting on tsetse."

The climate in the region where *Glossina morsitans* lives is characterized by a pronounced wet season, during which rainfall may exceed 70 or 80 inches, and a dry season, which may last for 6 or 7 months. A pupa may be likely to drown during the wet season and die from desiccation during the dry; drastic reductions in the numbers of *G. morsitans* may occur during either season. Jackson (1937), after having excluded the possibility of "density-dependent factors" concluded: "Meanwhile the writer is inclined to accept provisionally the view that the density of tsetse about Kakoma is being maintained at a mean level below that which the environment might allow; that the annual passage of a favourable season does not continue sufficiently long to bring the tsetse numbers up to saturation level before the ensuing unfavourable period drives them down again."

The biology of *Glossina tachinoides* is generally similar to that of *G. morsitans*, except that the former can feed on a wider variety of animals, including man. Its distribution, for this reason, is not likely to be restricted by the absence of suitable food. On the other hand, it has a characteristic behavior in hunting for food which, as we shall explain later, severely limits its distribution.

The northern boundary to the distribution of *Glossina tachinoides* tends to follow the isohyet for 30 inches of rain a year, which is associated with a dry season of 6 or 7 months. This species, taking advantage of the moisture and the vegetation associated with rivers, may penetrate into country where the annual rainfall is 16 inches and the dry season lasts for 8 months. The major projection of its distribution into the dry country shown in the northeast part of the map in Figure 13.18 is associated with the great Katagum River.

The southern limit to the distribution of *Glossina tachinoides* is not associated with any isohyet, although moisture is indirectly responsible for the limitation of its distribution in this direction also. This species has the peculiar habit of remaining within a few feet of the ground while hunting for food. They cannot fly through dense undergrowth, and they will not rise above any

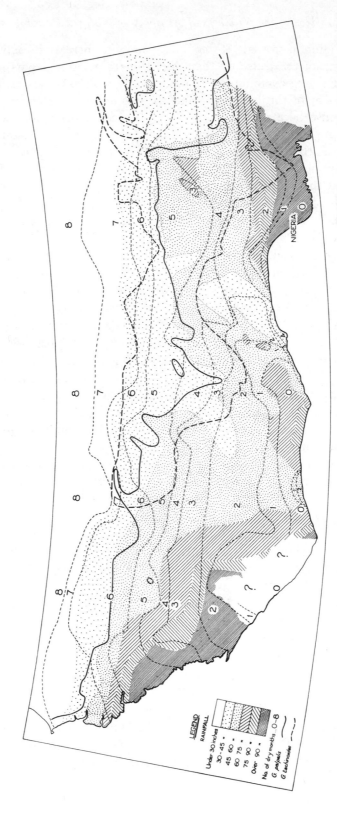

FIG. 13.18.—The distributions of *Glossina palpalis* and *G. tachinoides* in West Africa in relation to annual rainfall and the number of dry months in a year. (After Nash, 1948.)

growth that exceeds about 2 feet in height. Consequently, *G. tachinoides* are not found in areas of forest or woodland where the undergrowth is prominent. As a result, this species is excluded from some of the more humid regions which are inhabited by certain other species of *Glossina*, especially *G. palpalis*.

The northern limit to the distribution of *Glossina palpalis* follows the 45-inch isohyet fairly closely; in this region the dry season lasts for 5 or 6 months. In some places the distribution of this species extends into drier areas, but then it is restricted to the vegetation associated with rivers. In the south the distribution of *G. palpalis* extends to the coast, except that they are absent from the mangrove swamps and the swamp forests which form a fringe along the coast (Fig. 13.18); in areas which support these sorts of vegetation the soil is so wet that there are few places where a pupa is likely to escape drowning. Elsewhere *G. palpalis* is absent from local situations where the undergrowth is dense and continuous, for in these circumstances the flies cannot reach their food. Provided that there are some clearings which permit hunting, *G. palpalis*, in striking contrast with *G. tachinoides*, may colonize forest or woodland where the undergrowth is 6 feet high. Table 13.15 shows the striking

TABLE 13.15*

NUMBERS (AS PER CENT OF TOTALS) OF *Glossina palpalis* AND *G. tachinoides* WHICH SETTLED AT DIFFERENT LEVELS ON MAN STANDING UPRIGHT IN STREAM BED

HEIGHT ABOVE GROUND	LATTER PART OF DRY SEASON		SEASON OF HEAVY RAINS	
	G. palpalis	*G. tachinoides*	*G. palpalis*	*G. tachinoides*
Ground to ankle (0–4 inches).....	2	35	2	12
Ankle to knee (4–22 inches)......	13	44	23	54
Knee to waist (22–40 inches).....	24	18	32	21
Above waist (40–56 inches)......	61	3	43	13

* Data from Nash (1948).

differences in the habits of these two species with respect to the height above the ground at which they hunt: a man stood in the bed of a stream and counted the numbers of the two species which settled on him at several different levels The experiment was done twice, once late in the dry season and once during the season of heavy rains. In the latter part of the dry season 15 per cent of *G. palpalis* attacked below the knee, as compared with 79 per cent of *G. tachinoides*. During the wet season *G. tachinoides* fly even closer to the ground, and *G. palpalis* fly higher.

The distributions of four species of *Glossina* are shown broadly in Figure 13.19; they overlap without coinciding. In those areas where the distributions overlap, the resources which the different species need in common are never in short supply relative to the numbers of the flies; the species are not linked with one another through a common predator; nor do they interfere with one another's chance to survive and multiply in any other way that has been discovered Although they are all species of the one genus, they do not enter into one another's environments in any important way. The differences in their distributions are due, as we have seen, to differences in their physiology and behavior. According to current theory of speciation (chap. 16), these differences arose at some time in the past when the populations were isolated from each other.

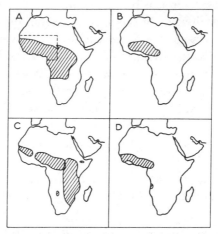

Fɪɢ. 13.19.—The distributions of four species of tsetse flies in Africa. *A, Glossina palpalis; B, G. tachinoides; C, G. morsitans,* showing eastern and western races; *D, G. longipalpis.* The dotted line in *A* shows the area which was mapped on a larger scale in Figs. 13.17 and 13.18. (After Gaschen, 1945; and Nash, 1948.)

13.34 *Fluctuations in the Numbers of Bobwhite Quail* Colinus virginianus *on 4,500 Acres of Farmland in Wisconsin*

Errington (1945) reported the results of 15 years' continuous study of a population of *Colinus virginianus* on 4,500 acres of farmland in Wisconsin. This is getting near the northern limit of the distribution of this species. Errington said: "To the northward are found only frontier populations of this 'farm-game' bird, usually sparse and discontinuous though sporadically abundant." What Errington called "cover conditions for the bobwhite" had long been deteriorating in this area, and it is likely that the deterioration continued slowly while this investigation was in progress. But what changes occurred were slight and gradual, and it was not considered necessary to take them into account.

The life-cycle and behavior of *Colinus virginianus* were mentioned in sections 10.322 and 12.31. The quail's habit of overwintering in coveys, the nature of the terrain, and the climate made it practicable, at any time during the winter, to count all the quail on the area. The routine varied during the 15 years, so that more information was available for some years than for others; but for every year it was recorded how many quails were present on the area at the beginning of the winter and how many at the end of the winter. These records are given in Figure 13.20, from which one can see at a glance the magnitude of the decreases that occurred during the winter and the increases that were made during the summer. With very few exceptions, the decrease during the winter was due to deaths among the local population. But the increase during the summer was not all due to breeding by the local population. At this time of the year the quail wandered freely over the countryside, and it was not possible to assess the gains and losses due to migration.

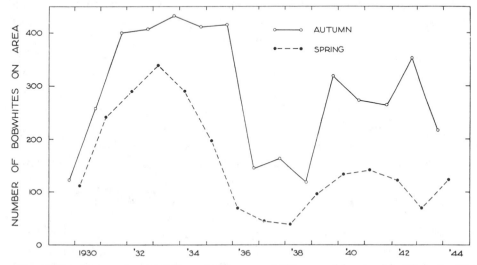

Fig. 13.20.—The numbers of *Colinus virginianus* on 4,500 acres of farmland in Wisconsin. The upper curve shows the numbers attained at the end of each summer; the appropriate comparisons between the upper and lower curves indicate the number of deaths during each winter and the increases which occurred during each summer. (After Errington, 1945.)

Most of the deaths which occurred during the winter could be attributed to starvation, exposure, or the activities of predators. The quail depended on seeds for food. The risk of starvation was greatest during a hard winter, when persistent snow or ice covered the ground and made their food inaccessible. For example, during the winter of 1935–36 there were "several weeks of the severest winter weather shown by Weather Bureau records for the region. Blizzard followed blizzard, with prolonged, intense cold and much snow. One after another of the Prairie du Sac coverts was nearly or quite depopulated of bobwhites. By spring, only 70 of the birds were left alive on the area." There had been 416 at the beginning of winter. But it was not necessary for the weather to be quite so extreme in order to kill many of the birds. The bobwhite starved if their food supply was cut off for a week or two. For example, the winter of 1937–38 "was also mild, but a food shortage aggravated by heavy snows in late January was attended by extreme losses. The area's spring level of 39 bobwhites remaining of an initial 163 was the lowest for which we have accurate measurement."

Deaths from exposure were less frequent and more difficult to assess than deaths from starvation. Errington summarized the position as follows: "In cold weather, the hunger-weakened are the likeliest to succumb. On rare occasions, sound birds in unsheltered places fall victim to cold, and others are imprisoned to starve or suffocate in hard-packed snowdrifts. In 1935/36 . . . bobwhites not only starved in the deep snow or died from the stress of the blizzards but also seemed to be worn down by the sustained cold."

The risks both from starvation and from exposure were accentuated by men who plowed the land during the autumn, thereby burying the seeds that might have served as food for the quail, and who slashed or burned "brush" which might otherwise have provided a place for the quail to seek shelter. Even if these activities did not result in an absolute shortage of food or an absolute shortage of hiding places, they might still increase the risks of starvation because the quail would not use food that was too far from a safe refuge. Or if they did, this would increase the risk of being taken by a predator.

The chief predator of the bobwhite quail was the great horned owl, *Bubo virginianus*. The numbers of these owls on the area varied from 4 to 8, with an average of 6. Other species of predators included a hawk, *Buteo jamaicensis,* gray fox, *Urocyon* sp.; red fox, *Vulpes* sp.; skunk, *Mephites* sp.; mink; and several sorts of weasels, *Mustela* spp., as well as dogs, cats, and men from the farms. During some winters the predators, among them, killed more than half the quail that were overwintering on the area. For example, there were 264 quail on the area at the beginning of the winter of 1941–42, but only 122 were still alive at the end of the winter. The weather was mild, and Errington attributed all the deaths to "non-emergency losses," which is nearly the same as saying that they were due to predators of one sort or another. On the other hand, there were several winters when the deaths attributed to predators were only 7 or 8 per cent. For example, there were 257 quail on the area at the beginning of the winter of 1930–31, and 236 were still alive at the end of the winter; all these deaths were attributed to predators, but they represented only 8 per cent of the numbers that were present at the beginning of the winter.

Since all the predators mentioned by Errington were species which would eat a variety of other foods besides the bobwhite quail and since most of them would, as a rule, tend to take that sort of food which was most plentiful, it was to be expected that the other animals which served as food for the predators might exercise an important influence on a quail's chance of being eaten. Errington called these "buffer species." They included mice, rabbits, and gallinaceous birds. The more likely ones were considered in more or less detail, but no relationship was found between their numbers and a quail's chance of being eaten. After exhausting all other likely possibilities, Errington concluded that the only component in a quail's environment which influenced the risks that it would be eaten by a predator was the number of animals of the same sort that were in the area. This led to the formulation of the "threshold concept," by which was meant that, for any particular area, there was a maximal number of quail which could live through the winter with a high degree of immunity from predators. If the numbers exceeded this "threshold," the supernumeraries would become "insecure" and would probably be eaten by predators. During the 15 years the "threshold" seemed to vary. Errington implied that this might be due to intrinsic changes in the behavior of the quail, but the evidence for this was not convincing (see below).

Figure 13.20 shows that not only did the number of deaths during the winter vary greatly but so did both the absolute and the relative increases during the summer. Scatter diagrams were constructed by plotting the relative increase during the summer against numbers present at the beginning of the summer. Two relationships were discovered in this way: (*a*) the relative increase during the summer was roughly inversely proportional to the number present at the beginning of the summer; (*b*) this relationship became more pronounced when the numbers of bobwhite quail and ring-necked pheasants were taken together. The trend emerged as a flat sigmoid curve. It was not defined precisely, because there was a wide scatter among the points.

We are now in a position to summarize the conclusions which may be drawn from the empirical facts discovered by this investigation: (*a*) The maximal number of quail which overwintered on the area was determined largely by the behavior of the quail themselves in relation to the distribution and abundance of suitable places for them to live (sec. 12.31). (*b*) The minimal numbers were determined largely by the weather; cold weather was directly responsible for some deaths, but more important was the indirect influence of snow and ice in reducing the stock of accessible food. (*c*) The numbers attained by the end of the breeding season depended partly on the numbers present at the beginning of the summer and partly on the relative increase during the breeding season. (*d*) The relative increase during the breeding season depended partly on the numbers of quail and partly on the numbers of a nonpredator (the ring-necked pheasant) which required to share some of the same resources and also probably interfered with the quail in a more subtle way as well. (*e*) There were runs of years in which the quail were plentiful, followed by runs of years in which they were scarce; but the sequence of 15 years' records was too short to permit a statistical test of any hypothesis about cycles or periodicity.

Errington carried the analysis of these data further than this. He analyzed them more especially in terms of the two hypotheses which we mentioned earlier, namely, the "threshold concept" in relation to the rate of survival during winter and the "inverse density law" in relation to the relative increase during summer. He constructed curves to represent these hypotheses and found that whereas most points fell approximately along the trend lines, there were some which deviated widely from the trend defined by the majority. Moreover, there was a relationship between the points which deviated most from the major trend in each case. This is best expressed by saying that a winter in which the number of deaths from predators was unexpectedly low was likely to follow, or be followed by, a summer in which the relative increase was also unexpectedly low. It was suggested that this might be explained by assuming that this population of *Colinus virginianus* experienced "depression phases" analogous to those supposed to be manifested by the populations of certain species of vertebrates (*Lepus, Microtus, Lynx,* and certain birds and fishes) in which the fluctuations are said to be periodic (sec. 13.4).

These data do not in themselves provide convincing evidence for the reality of "depression phases." Errington preferred this explanation because it also fits in with similar observations on populations of other species, especially the musk-rat *Ondatra zibethicus.* It often happens that unusually low birth-rates in populations of muskrats seem to synchronize with low rates of increase not only for the same species in different areas but also for quite unrelated species such as hares or grouse throughout the country. There seems to be no way of explaining these observations in terms of the ordinary components of environment. Nor do they seem to be due to the innate qualities of the animals. In the present state of our knowledge the term "depression phase" may be used to describe the phenomenon without indicating any explanation for it. The idea of "depression phases" has been linked with the idea of "periodicity" in the fluctuations in numbers in certain vertebrates. But one need not necessarily depend on the other.

13.4　THE PHENOMENON OF "PERIODICITY" IN THE NUMBERS OF CERTAIN VERTEBRATES

Elton and Nicholson (1942*b*) consulted the records of the Hudson's Bay Company and compiled tables showing the relative number of furs of the lynx *Lynx canadensis* received by the company each year from 1736 to 1934. They divided the total area into regions. Moran (1953) analyzed the records from the Mackenzie River region for the period 1821–1934. He chose them because this was the only region from which "complete figures are available over a long period of years." Moran also pointed out: "When the total captures for the whole area are considered for each year, a rather blurred picture of the cycle appears, but on splitting up the records into the individual regions, the cyclic behaviour of the population in each region . . . becomes quite clear."

We reproduced these records in Figure 13.21, *B,* because they are probably the best example of the "10-year cycle" that can be found in the literature. It has been said that the numbers of the lynx depend on the numbers of the snowshoe hare, which is its staple food. The numbers for the snowshoe hare

Lepus americanus, which were used in Figure 13.21, *A*, were copied from a small diagram in Dymond (1947), because the original records (Hewitt, 1921) were not available to us. Figure 13.22 shows the number of furs of muskrats *Ondatra zibethica* received by the Hudson's Bay Company (Elton and Nichol-

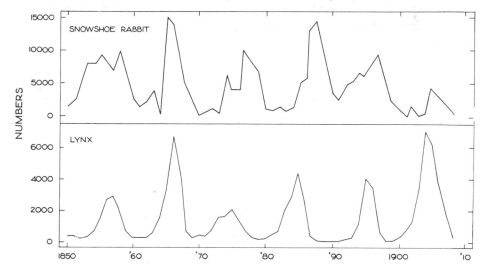

FIG. 13.21.—Fluctuations in the numbers of snowshoe rabbits (total area ?) and the lynx (Mackenzie River region), as indicated by records of pelts kept by the Hudson's Bay Company. (After Hewitt, 1921; and Elton and Nicholson, 1942*b*.)

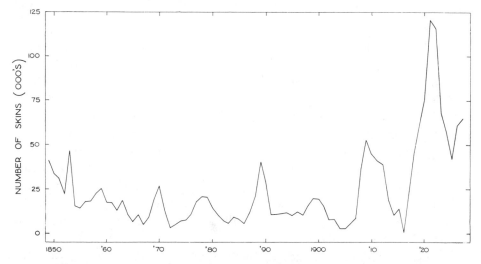

FIG. 13.22.—Fluctuations in the numbers of muskrats *Ondatra zibethica* for the Mackenzie River region, as indicated by the records of the Hudson's Bay Company. (After Elton and Nicholson, 1942*a*.)

son, 1942*a*). Figure 13.23 shows the numbers of red grouse *Lagopus scoticus* shot by sportsmen on three moors in Scotland between 1850 and 1945 (Mackenzie, 1952). The last two series have also been said to provide evidence for "cycles," but they are obviously not so "regular" as that of the lynx, which seems outstanding in this regard. Most writers seem to accept the reality of cycles without question. Nevertheless, the evidence warrants a critical examination.

Two aspects of "cycles" have attracted attention, namely, the amplitude of the fluctuations and their periodicity. For example, Dymond (1947) stated

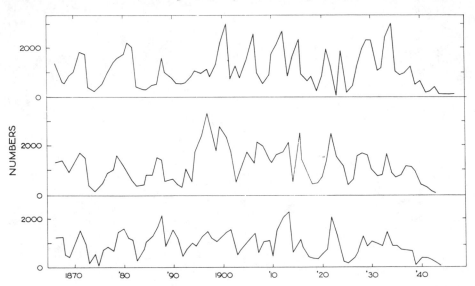

Fig. 13.23.—Fluctuations in the numbers of red grouse shot on three moors in Scotland. (After Mackenzie, 1952.)

that the maximal numbers of *Lynx canadensis* exceeded the minimal numbers in the ratio of 100 to 1 in the northern part of British Columbia, 50 to 1 in the middle part of the province, and 20 to 1 farther south. The number of salmon in the Fraser River, British Columbia, were estimated to have fluctuated between the extremes of 15,000 in 1907 and 240,000 in 1913. These are large fluctuations, and doubtless they have seemed more impressive because of the commercial value of most of the species concerned. But these fluctuations may be small compared with those which occur in natural populations of locusts, grasshoppers, thrips, and many other species of invertebrates. In other words, the muskrat, the lynx, the snowshoe hare, the grouse, and the salmon are not unusual with respect to the amplitude of the fluctuations which are found in their numbers.

A truly cyclic variable, such as the sine or cosine of an angle, waxes and wanes with constant amplitude and phase. A variable may still be properly called "cyclical" if it is determined by two components, one of which is truly cyclical and the other added to it is random. A variable may be properly called "oscillatory" if the amplitude and phase of the fluctuations about the mean (or trend line) are not random with respect to their order along the abscissae. The usual test for randomness in a series is to calculate the serial correlation coefficient (Kendall, 1948, p. 402). The difference between the two sorts of series was summarized by Kendall (1948, p. 398):

> In a cyclical series the maxima and minima, apart from disturbances due to the super-position of a random element, occur at equal intervals of time and are therefore predictable for a long way into the future—for so long, in fact, as the constitution of the system remains unchanged. In oscillatory series, on the other hand, the distances from peak to peak, trough to trough or upcross to upcross, are not equal, but vary very considerably. Similarly, in the oscillatory series the amplitudes of the movements may vary very substantially, whereas in a cyclical series they should be constant (again, except in so far as superposed random elements disturb them) . . . the time-series observed in practice are very rarely cyclical as we have defined the term. . . . The far more usual case is that of varying amplitude and period from peak to peak or upcross to upcross.

The series plotted in Figures 13.21, 13.22, and 13.23 are clearly not cyclical, because the variation in the amplitude of the oscillations is far too great. On the other hand, Moran (1952, 1953) showed by the method of serial correlation that the series represented by the numbers of *Lynx canadensis* and *Lagopus scoticus* were significantly nonrandom with respect to years; and there need be little doubt that the same could be shown for the muskrat (Fig. 13.22). Although they are not cyclical in the strict mathematical meaning of the word (because the amplitude is not constant), these series may be periodic if the turning points recur at regular intervals.

If the minor fluctuations in the curves shown in Figures 13.22 and 13.23 are ignored, the more obvious periods of maximal numbers seem to be repeated at approximately equal intervals. But a dilemma is encountered when one tries to measure the intervals between "peaks" precisely: the regularity largely disappears if the maxima are determined objectively by counting as a maximum every point that is greater than the ones that come immediately before and after it; on the other hand, if the minor fluctuations are ignored and only the big peaks are counted, there is no objective way of measuring the precise intervals between maxima: every maximum and minimum has to be fixed arbitrarily. It is not permissible to smooth the raw data, because an oscillatory series may be generated from a random one by taking moving averages and by certain other devices which may be used for smoothing (Kendall, 1948). The information in Table 13.16 was compiled by Cole (1951) by counting every turning point; the information in Table 13.17 was gathered by Dymond (1947), using arbitrary methods of ascertaining the periods between "big peaks." A number of species occur in both tables, and the discrepancies are striking. Let us follow up Cole's approach first.

If we take many numbers from a table of random numbers and plot them in the order that they come from the tables, they will fluctuate about a horizontal line drawn through their arithmetic mean. Wherever there is a number which

TABLE 13.16
INTERVALS BETWEEN MAXIMA IN NUMBERS OF 8 SPECIES OF VERTEBRATES ESTIMATED OBJECTIVELY
BY METHOD OF COLE

Species	Country	No. of Oscilla- tions	Mean Interval (Years)	C.V. (Per Cent)	P
Lynx*	Canada	23	5.82	50.1	< 0.01
Lynx	Finland exc. S.W.	93	3.56	41.5	< .01
Field hare	England	48	3.60	48.1	< .01
Rabbit	England	58	3.53	47.5	< .01
Partridge	England	117	3.21	39.8	< .01
Arctic fox*	Labrador	124	3.50	27.1	Nonsig.
Red fox*	Labrador	124	3.37	31.1	Nonsig.
Red fox	Finland	93	3.66	40.0	< 0.01
Random numbers	. . . .	. . .	3.0	37.3	. . . . .

* The species marked by an asterisk feature also in Table 13.17. The last column gives the significance of the differences between the coefficients of variation for the empirical series and the random numbers. Data from Cole (1951).

TABLE 13.17
INTERVALS BETWEEN "BIG PEAKS" IN NUMBERS OF 22 SPECIES OF VERTEBRATES IN CANADA

SPECIES		USUAL INTERVAL (YEARS)	EXTREME RANGE (YEARS)
Common Name	Technical Name		
Arctic fox*	*Alopex lagopus*	4	3– 6
Red fox*	*Vulpes fulva*	9–10	8–13
Lynx*	*Lynx canadensis*	9–10	7–12
Snowshoe rabbit	*Lepus americanus*	9–10	8–11
Muskrat	*Ondatra zibethica*	9–10	?8–?12
Field mouse	*Microtus pennsylvanicus*	4	. . . .
Lemming	*Lemmus trimucronatus*	4	. . . .
Marten	*Martes americanus*	9–10	8–11
Fisher	*M. pennanti*	9–10	8–11
Mink	*Mustela vison*	9–10	6–12
Goshawk	*Accipiter gentilis*	9–10	. . . .
Rough-legged hawk	*Buteo lagopus*	4	3–5
Ruffled grouse	*Bonasa umbellus*	9–10	9–12
Willow ptarmigan	*Lagopus lagopus*	9–10	. . . .
Sharp-tailed grouse	*Pediocetes phasianellus*	9–10	. . . .
Horned owl	*Bubo virginianus*	9–11	. . . .
Snowy owl	*Nyctea scandiaca*	4	?3–?6
Great northern shrike	*Lanuis excubitor*	4	?3–?6
Pine grosbeak	*Pinicola enucleator*	5–6	. . . .
Canadian salmon	*Salmo salar*	9–10	. . . .
Sockeye salmon	*Oncorhynchus nerka*	4	. . . .
Pink salmon	*O. gorbuscha*	2	. . . .

* The species marked by an asterisk also occur in Table 13.16. After Dymond (1947).

is larger than the ones on either side of it, there will be a maximum in the
curve; wherever there is a number smaller than the ones on either side of it,
there will be a minimum in the curve. Half the turning points in the curve will
be maxima. Now it is possible, by making use of the algebraic method of
permutations and combinations, to calculate the probability that maxima will
be separated by 2, 3, 4, . . ., n, places. Cole (1951) did this and obtained the
theoretical distribution shown in the second column of Table 13.18; the distribution shown in the third column was calculated by Cole from Schulman's
(1948) records for the serial distribution of growth-rings in the Douglas fir;

TABLE 13.18

FREQUENCY-DISTRIBUTIONS OF INTERVALS BETWEEN MAXIMA FOR ANY LONG
SERIES OF RANDOM NUMBERS AND EMPIRICAL SERIES BASED ON SIZE OF
GROWTH-RINGS IN DOUGLAS FIR

INTERVAL BETWEEN MAXIMA (ORDINAL UNITS FOR RANDOM NUMBERS; YEARS FOR DOUGLAS FIR)	FREQUENCY OF INTERVALS OF SPECIFIED DURATION	
	Random Numbers (Per Cent)	Douglas Fir (Per Cent)
2..	39.1	37.2
3..	34.4	32.9
4..	17.4	19.4
5..	6.5	7.7
6..	1.9	1.3
7..	0.5	0.8
8 and more...........................	0.2	0.6
Mean interval.......................	3.0*	3.07
Coefficient of variation (per cent).......	37.3*	37.3

* The numbers marked with an asterisk are repeated in Table 13.16. After Cole (1951).

any ring which was larger than the ones on either side of it was counted as a maximum. The distribution was calculated from 504 intervals. There is close agreement between the two distributions, and Cole concluded that in the Douglas fir growth-rings of varying sizes occurred at random with respect to time.

The mean and the coefficient of variation for the theoretical series are given at the foot of Table 13.16. We have calculated the significance of the differences between the coefficients of variation for the empirical series and the series of random numbers. The means of the empirical series are of the same order as that of the theoretical one, but the coefficients of variation are significantly larger for 6 of the 8 species listed in Table 13.16. This leads to the surprising conclusion that three-quarters of the empirical series are more variable than the theoretical one. None was significantly less variable than the theoretical series, which they would need to be if they were to provide evidence for periodicity. The departures from randomness (which may also be demonstrated by the method of serial correlation) happen, not as has so often been thought, because the maxima recur with a regularity greater than would be expected by random chance, but rather because the maxima recur with less regularity than they would in a random series. In other words, the curves representing these populations are oscillatory but not periodic.

This conclusion is analogous to the one reached in section 13.02. There we showed by the appropriate statistical tests that, for many species of animals, the individuals in natural populations are distributed nonrandomly with respect to space and that the departures from randomness are due to excessive "patchiness" or irregularity in the distributions. Here we have to conclude with Cole (1951, 1954) that (for many species of vertebrates at least) the individuals are distributed nonrandomly with respect to time and that the departures from randomness are due to excessive irregularity (or lack of periodicity) in the phase of the oscillations.

The discrepancies between Table 13.16 and 13.17 for species that are common to both may be due partly to the methods of ascertaining the numbers, as well as to the different methods of measuring the intervals between maxima. Figure 13.21, *B*, refers to the lynx in the Mackenzie River district; we have calculated the mean interval between maxima as 8.1 years, with a coefficient of variability of 31.5 per cent. This differs from the figures given by both Cole and Dymond, which refer to lynx in a wider area. There is always a dilemma to be faced in gathering the raw material for the study of "cycles": in order to have a sufficiently long series of figures, it is necessary to go back a long time into the past; the farther one goes back, the more doubtful are the figures. Possible variations in the area concerned, in the number of trappers, in their skills and objectives, and in a variety of other unknowns combine to throw doubt on the results. The numbers plotted in Figure 13.21, *B*, differ from those published by earlier authors; Elton and Nicholson (1942*b*) went thoroughly into the reasons for departing from the earlier publications.

The series in Figure 13.21, *B*, is definitely nonrandom, because Moran (1953) has shown that the serial correlation coefficient is large and significant. Compared with a random series, the phase and the amplitude of the oscillations are large, but the variability in the phase is only slightly less than that of a random series. This series is therefore markedly oscillatory (nonrandom), but it does not provide strong evidence for periodicity—especially when it is recalled that this series was arbitrarily selected because it was probably the most "cyclical" one to be found in the literature.

It might be argued that Cole's approach, while having the merits of objectivity and precision, lacks reality and that the figures in Table 13.17, though they depend on subjective judgment, are nevertheless more real than those in Table 13.16. This may be so, but there is no way to put the matter to the test. It might also be suggested that since the methods of ascertaining the numbers are so uncertain, it is hardly worth while to apply rigorous methods to their analysis. In this connection it would be well to recall Kendall's (1948, p. 402) warning: "Experience seems to indicate that few things are more likely to mislead in the theory of oscillatory series than attempts to determine the nature of the oscillatory movement by mere contemplation of the series itself."

Cole (1951) considered the arguments in favor of counting every turning point instead of only the big ones. Apart from the obvious statistical reasons for this choice, there are two strong biological arguments against ignoring the minor fluctuations: (*a*) A level of favorableness, operating for a certain period, in the environments of individuals in a small population may result in a small absolute change in numbers, whereas a large population in identical circumstances (except for its own density) may experience a great change in its numbers, measured absolutely. Cole likened this to a small and a large bank balance increasing at compound interest. (*b*) An animal's chance to survive and multiply (and hence the rate of change in the population) is determined by various components of the environment, which are themselves distributed continuously (in the statistical sense). Small and large variations in the environmental components will occur with a frequency which is determined by the means and the variances of the distributions; certainly, these will be reflected in small and large variations in the numbers of the animals. There is just no reason at all for attributing greater reality to the big variations than to the small. We were able to demonstrate this principle quantitatively with respect to *Thrips imaginis* (sec. 13.11). The ecology of *Lepus, Lynx, Vulpes, Lagopus,* etc., may be more complex than that of *T. imaginis;* but the same principles may be expected to hold with them also.

Our success in relating the numbers of *Thrips imaginis* quantitatively to certain environmental components depended largely on three conditions: (*a*) we had adequate knowledge about the physiology and behavior of *T. imaginis;* (*b*) we were working in an area where the geography and especially the climatology had been investigated thoroughly; and (*c*) our records of the relative numbers of *T. imaginis* were adequate with respect to both the number of samples and the precision with which they were taken.

The records on which the discussions in this section have been based are inferior to those for *Thrips imaginis* with respect to all three conditions, especially the first and the last: the number of pelts received by the Hudson's Bay Company or the number of cases of salmon packed along the Fraser River must depend on many things besides the number of animals in the forests or the river. In studying the ecology of these species, as with all others, it is best to set out to explain all the variability in the recorded numbers and to take into account all the environmental components that seem relevant. We reached the same conclusion after discussing the ecology of the bobwhite quail (sec. 13.34). The investigation is more likely to be successful if it is firmly based on a sound knowledge of the biology of the animal and the geography of the area where it is living. The long series of "records" for some of these species may be alluring; but, in the future, progress may depend on getting more precise information about the animals themselves and their environments.

CHAPTER 14

A General Theory of the Numbers
of Animals in Natural Populations

*From my early youth I have had the strongest desire to understand or explain
whatever I observed,—that is, to group all facts under some general laws. These
causes combined have given me the patience to reflect or ponder for any number of
years over any unexplained problem. As far as I can judge, I am not apt to follow
blindly the lead of other men. I have steadily endeavoured to keep my mind free so
as to give up any hypothesis, however much beloved (and I cannot resist forming
one on every subject), as soon as facts are shown to be opposed to it.*

DARWIN, *Autobiography*

14.0 INTRODUCTION

IN THIS chapter we have to build a general theory about the distribution and
numbers of animals in nature. This should summarize in general terms as
many as possible of the facts which we have discussed in the empirical part of
the book and so provide a general answer to the questions we propounded in
section 1.1: Why does this animal inhabit so much and no more of the earth?
Why is it abundant in some parts of its distribution and rare in others?

Elton (1949, p. 19) wrote: "It is becoming increasingly understood by population ecologists that the control of populations, i.e., ultimate upper and lower
limits set to increase, is brought about by density-dependent factors either
within the species or between species (see Solomon, 1949). The chief density-
dependent factors are intraspecific competition for resources, space or prestige;
and interspecific competition, predators or parasites; with other factors affecting the exact intensity and level of these processes." Elton said quite precisely
what he meant by "density-dependent factors"; in section 2.12 we discussed
some of the other meanings which have been attributed to the same phrase;
like many other ecological terms, this one lost most of its strength shortly
after it was coined. We hope that Elton overstated the case. We believe that
it would be nearer the mark to say that the various assertions about "density-
dependent factors" and "competition" which are familiar to ecologists are
just about the only generalizations available in this field. The student of
ecology may either accept or reject them. Hitherto if he rejected them, he has
had nothing to put in their place.

The statement that the ultimate upper and lower limits set to increase in a population can be determined only by "density-dependent factors" may be taken as axiomatic for the highly idealized hypothetical animals of the sort which may be represented by the symbols in a simple mathematical model. In this case the limits which are referred to are theoretical quantities which must be deduced by mathematical argument. They cannot, by their very nature, be related to the empirical quantities which are got by counting the numbers of animals in natural populations. There are two chief reasons for this: (*a*) the idealized hypothetical populations are very different from what is actually found in nature, and (*b*) one would not expect to come across a limiting density in any finite number of observations.

Yet this mistake is commonly made, as Elton has pointed out. The usual generalizations about "density-dependent factors," when they refer to natural populations, have a peculiar logical status. They are not a general theory, because, as we have seen, especially in chapter 13, they do not describe any substantial body of empirical facts. Nor are they usually put forward as a hypothesis to be tested by experiment and discarded if they prove inconsistent with empirical fact. On the contrary, they are usually asserted as if their truth were axiomatic. A good example of this approach is seen in the passage from Smith (1935; quoted in sec. 2.12), where he argued that since weather is known to "regulate" the numbers of animals, it must therefore be a density-dependent factor. These generalizations about "density-dependent factors" and competition in so far as they refer to natural populations are neither theory nor hypothesis but dogma.

We often find the expressions "balance," "steady-density," "control," and "regulate" used in theoretical discussions of populations. Their meanings may be obscure, especially when they are used in relation to natural populations. They stem from the dogma of "density-dependent factors," and they are allegorical. Unless their meanings are made very clear, it is best to be cautious about any passage in which they are used.

Our theory is not concerned with these rather allegorical properties of populations but with numbers that can be counted in nature. In each of the sections which follow we first state the principles of the theory in general terms and with the aid of simple diagrams. Then we refer back to the natural populations which have been discussed in earlier chapters and show how particular empirical observations may be fitted into the general theory; for the theory may be regarded as sound only if it serves to explain all, or most, of the empirical observations that may be brought forward to test it.

14.1 COMMONNESS AND RARENESS

If we say that species A is *rare* or that species B is *common*, we can only mean that individuals of species A are few relative to some other quantity which we can measure, and individuals of species B are numerous relative to the same or some other quantity. We might arbitrarily choose a number relative to a certain area and decide to call a species rare if they are, say, fewer than one (or one million) individuals per square mile. This approach may be necessary for the hunter or fisherman, but it does not, at this stage, help us to think clearly about the theory of ecology. So we shall put it aside for the present.

Another meaning for this sentence might be that in a certain area individuals of species A are few relative to the individuals of species B. Certain mathematical expressions have been developed to describe the relative numbers of different species found in large communities or large samples from communities. The size of the community is measured by the number of individuals of all species that are counted. Williams (1944) gave the name "index of diversity" to the coefficient $\alpha = n_1/x$ derived from the series

$$n_1, \frac{n_1}{2}x, \frac{n_1}{3}x^2, \frac{n_1}{4}x^3, \ldots, \text{etc.,}$$

where n_1 is the number of species represented in the sample by one individual and x is a positive number less than 1. Large values of α indicate diversity in the community. Preston (1948) fitted a "truncated" normal curve to the frequencies of species represented in the sample by 0–1, 1–2, 3–4, 5–8, 9–16, $\ldots$, $(2^{n-1} + 1)$–2^n, individuals per species. In this expression the diversity of the community is measured by the mean and the variance of the normal curve and the position of the "truncation."

Both expressions depend on the assumption that in any large community there will be a few species represented by many individuals and many species represented by few individuals. This assumption is confirmed by the goodness of the fit obtained when the theoretical curves are tested against empirical records. In other words, most communities seem to be dominated by a few species. This may be an important fact for students of "community ecology." It is, indeed, on facts of this sort that they base their theory of "dominance" and other important theories about communities and the relationships of the species which constitute them. But this way of looking at commonness and rareness and the studies which stem from it have little to contribute to the general theory of ecology, using "ecology" in the narrow sense defined in section 1.0. So we say no more about them.

A third way of considering commonness and rareness is to relate the number of individuals in the population to the quantities of necessary resources, food, places for nests, etc., in the area that it inhabits. This is the common-sense way for the theoretical ecologist and the practical farmer to look at the matter. The farmer is not interested in whether the caterpillars in his crop of wheat are more or less numerous than, say, the mites which eat the dead grass on the surface of the soil. He wants to know whether the caterpillars are numerous enough to eat much of his crop. The ecologist has the same point of view, but his interest goes deeper. If the caterpillars are so few that they eat only a small proportion of the stock of food available to them, the ecologist has to inquire what other environmental components may be checking their increase.

14.11   *The Conditions of Commonness in Local Populations*

A population of a certain species, living in a certain area, may consistently use all the stocks of a particular resource, say food or places for nesting, that

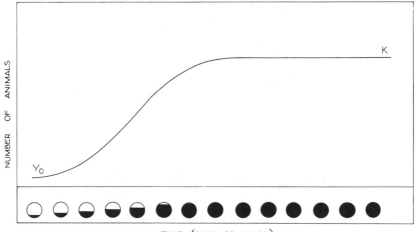

TIME (DAYS OR YEARS)

Fig. 14.01.—The growth of a "local population" (i.e., the population in a "locality") whose numbers are limited by the stock of some nonexpendable resource, such as nesting sites. The initial numbers (i.e., the number of immigrants who colonize the locality) are represented by Y_0; the maximal numbers that the resources of the locality will support are represented by K. The circles repeat the information given by the curve. The proportion of the circle shaded represents the number of animals at the specific time as a proportion of the maximum.

occur in the area. The simplest case of this, though perhaps the most unusual in nature, may be found when there is no interaction between the resource and "other animals of the same kind." This happens when the animals can use the resource without destroying it. Such was the case in the imaginary example we gave in section 2.121. The bees in that example did not destroy their nesting places by using them. There was the same number of holes for nest-building generation after generation, irrespective of the numbers of bees in each generation. If nothing else checked the bees, their numbers would be determined entirely by the number of holes for nests. We have illustrated this principle in Figure 14.01. The asymptote for the curve represents the number of bees in a local situation when all the nesting places are used. The shaded parts of the circles represent the population as a proportion of what the total resources of the area will support. A natural population which comes very close to the simplicity of the imaginary one is the population of great tits mentioned by Kluijver (sec. 12.31) as living in a wood where there was a shortage of holes for building nests. Another one, which is nearly as simple, is the one described by Flanders (sec. 12.23) for *Metaphycus helvolus* living on oleanders. In this case living food in the form of migrating scale insects was the resource in short supply. It was fed into the population of *Metaphycus* at a

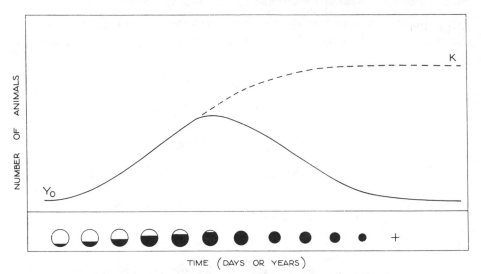

F$_{IG}$. 14.02.—The solid line represents the growth of a local population whose numbers are limited by a diminishing resource, such as living food, which becomes less, the more animals there are feeding on it. The symbols have the same meanings as in Fig. 14.01; the circles grow smaller because the plant (or population of prey) grows smaller and eventually dies out because there are too many animals eating it. The cross indicates that this is the place where a local population recently became extinct (see Figs. 14.07 and 14.08).

fairly steady rate, which was largely independent of the numbers of animals waiting to eat it. But there were always enough *Metaphycus* to eat all the food that was provided.

When living food is the resource that is fully used up, there may be certain complicated interactions between "food" and "other animals of the same sort"; the result may be a reduction not only in the total stocks of food but also in the proportion of it that is effectively used (sec. 11.22). The sequence of events usually culminates in the complete destruction of localized stocks of food. This accentuates the patchiness of the distribution of food; and the animal's chance of finding food comes increasingly to depend on its powers of dispersal. A characteristic sequence of events was described for *Cactoblastis cactorum* in section 5.0. The general case is illustrated in Figure 14.02. A horizontal line drawn through K would represent the maximal number of animals that could live in this locality if they used up all their resources. The curve represents the rise and fall in the number of animals from the time that they first find the place to the time that they die out from lack of food. The diminution in the amount of food and its ultimate extinction from this place are indicated by the diminishing size of the circles from left to right. The cross indicates a place from which a local population has recently died out. The food becomes less and ultimately disappears, because there are too many animals feeding on it. We mentioned *C. cactorum* feeding on *Opuntia* spp. as an example of this (sec. 5.0); *Ptychomyia remota* is a carnivore for which the same sequence of events was described in section 10.33; other examples were given in section 10.321; we can think of many more examples from carnivorous species than from herbivorous ones. It is unusual to find herbivorous animals eating out their stocks of food, even in local situations. Animals which are living in the circumstances which we have described in this section must be counted as common, no matter how few they may be per square mile. Conversely, animals which are living in the circumstances which we describe in section 14.12 must be counted as rare with respect to their stocks of food, etc., no matter how many of them there may be per square mile.

14.12 *The Conditions of Rareness in Local Populations*

Very few of the natural populations which were described in chapter 13 ever became numerous enough to make use of all their stocks of food, etc. Most natural populations are like this. The numbers fluctuate, perhaps widely; but they do not become numerous enough, even during periods of maximal abundance, to use more than a small proportion of their resources of food, nesting sites, and so on.

The general case for this condition is illustrated in Figure 14.03. The curves and the symbols have the same meanings as in Figures 14.01 and 14.02; note that the size of the circles remains the same from start to finish, because the stocks of food, etc., are not appreciably reduced by the activities of the animals; the circles, of course, never become completely black. The circle with a cross in it provides an alternative to the one above it. These alternatives indicate that the population in a locality runs a risk of being extinguished but also has a chance to survive as a small remnant. In the latter case the population may increase again when circumstances become favorable once more.

The numbers continue to increase while births exceed deaths, that is, while r remains positive; they begin to decrease when r becomes negative. A large part of this book has been devoted to a discussion of the ways that environment may influence the three components of r—fecundity, speed of development, and duration of life. We do not need to reiterate here that any environmental component may have its appropriate influence and that, at any one time, some may be more influential than others. We single out weather and discuss it in section 14.13, partly because its influence is more subtle than that of the other components, and partly because this is a subject that has been much misunderstood.

In section 14.2 we discuss, in relation to the general theory, the way that weather, food, predators, and a place in which to live may influence the value of r and hence determine the numbers in natural populations which occupy substantial areas and which may, or may not, be short of some necessary re-

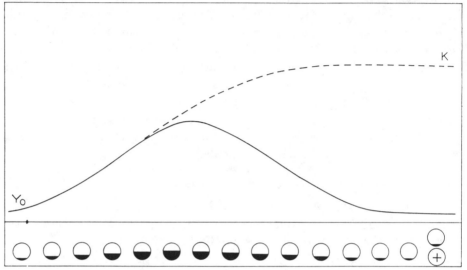

TIME (DAYS OR YEARS)

FIG. 14.03.—The solid curve represents the growth of a local population which never uses up all the resources of food, etc., in the locality because its numbers are kept, by weather, predators, food (in the relative sense of sec. 11.12), or some other environmental component, at a level well below the maximum that the resources of the locality could support. The symbols have the same meanings as in Figs. 14.01 and 14.02. The two symbols at the end indicate alternative conclusions to the history of this local population on this occasion: it may be extinguished (*circle with a cross in it*), or a remnant may persist, perhaps to increase again when circumstances change (*circle with a remnant of shading*).

source. We choose these components because they are often important in nature. The reader can readily fit other environmental components into the general theory.

Figure 14.03 refers generally to any population which is not short of any resource, no matter whether r is chiefly influenced by weather, food, predators, or some other environmental component. But it lacks generality in one respect: it does not cover the special case of the population in which a more or less constant proportion is sheltered from a danger which is likely to destroy all those that are not so sheltered. A good example of this was provided by the artificial population of *Ephestia*, which was kept in check by a predator (sec. 10.222). A natural example, which was nearly as good, was described by Flanders for a population of scale insects *Saissetia oleae*, living on the shaded parts of oleander (sec. 12.23). This exception to the general rule is covered by Figure 14.04. The meanings of the curves and symbols are self-evident.

14.13 *The Way in Which Weather May Determine Commonness or Rareness in Local Populations*

We build up our theory about weather from a simple beginning with Figure 14.05. The abscissae are in units of time—days, years, or generations. The

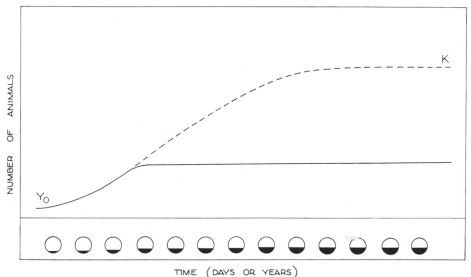

FIG. 14.04.—The solid curve represents the growth of a local population in which a more or less constant number of the individuals are sheltered from some danger which destroys all those which are not so sheltered. This number is not enough to use up more than a small proportion of the total resources of food, etc., in the locality.

ordinates for each curve are numbers of animals; K represents the number of animals which the area could support if the total resources of food were fully used up. The number indicated by Y_0 is the remnant left when the unfavorable period ends and is also the nucleus for multiplication during the next favorable period. The rate of increase which pertains during the favorable period is called r.

Let us suppose that the three pairs of curves represent three ways in which the weather may determine the number of animals in a local population. The curves on the left relate to a place where the weather is favorable, and those on the right relate to a place where the weather is severe. In the top pair the two curves start from the same level at the beginning of the favorable period (Y_0 is constant); and they rise at the same rate because r is constant. But the curve for area A rises farther because the favorable period lasts longer. With the middle pair, the two curves rise at the same rate (r is constant); they continue rising for the same interval of time (favorable period, t, is constant). But the curve for area A rises farther than that for B because it started from a higher level (Y_0 is greater for A). This means that in A the unfavorable period was shorter or else the catastrophe that caused it was less severe, or for some other reasons the population was still relatively large when the weather

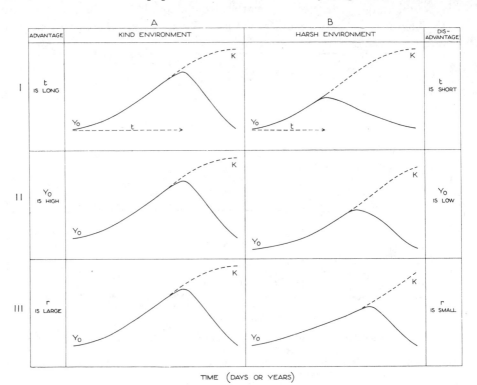

TIME (DAYS OR YEARS)

Fig. 14.05.—Three ways in which weather may influence the average numbers of animals in a locality. The numbers in the locality are increasing during spells of favorable weather and decreasing during spells of unfavorable weather. Two areas A and B are compared with respect to three qualities. Quality I determines the duration of the favorable period; quality II determines the severity of the unfavorable period; and quality III determines the rate of increase of the population during the favorable period. The numbers which would be attained if all the resources of food, etc., in the area were made use of are indicated by K; the numbers to which the population declines during the unfavorable period is represented by Y_0. For further explanation see text.

changed and allowed the animals to start increasing again. With the bottom pair, the two curves start from the same level (Y_0 is constant); they continue rising for the same interval of time (favorable period, t, is constant). But the curve in area A rises higher because it is steeper (rate of increase, r, is greater for A). Taking all three curves into account, one can easily see that the animals would be more numerous, on the average, in area A than in area B. This principle has been stated in completely general terms. The model which we describe in section 1.1 provides a particular example of it.

14.2 THE PRINCIPLES GOVERNING THE NUMBERS OF ANIMALS IN
NATURAL POPULATIONS

Each of Figures 14.01–14.05 represents the trend in numbers in one locality. A natural population occupying any considerable area will be made up of a number of such local populations or colonies. In different localities the trends may be going in different directions at the same time. It is therefore feasible to represent the condition of the population in a large area by drawing a col-

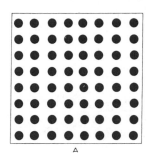 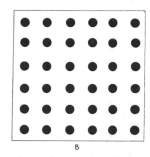

FIG. 14.06.—The populations of large areas are made up of a number of local populations. In the two areas which are compared in this diagram the resources of every locality are fully used up (as in Fig. 14.01), but there are more favorable localities in area *A* than in area *B*.

lection of circles, each one of which represents the condition in a local population. Since we are now considering larger areas, we must take into account, in addition to the principles set out in section 14.1, the dispersive powers of the animals and their food. Also we are now in a position to widen the meaning in which we use "common" or "rare" so that it includes the number of animals in the population relative to the area over which they are distributed (sec. 14.0). We do this in the examples which follow.

To start with the simplest case first, we make up an example directly from Figure 14.01. Let us suppose that in Figure 14.06 area *A* includes many localities where fenceposts have many holes suitable for the bees to build their nests in. And suppose that area *B* is like *A* in every respect except that there are few localities where the fenceposts carry many holes that are suitable for nests. Every hole will be used. All the circles are completely blackened in both areas. The two areas are alike, in that the bees are equally common with respect to their stocks of nesting sites. But the areas are different, in that the bees will be more numerous in *A* than in *B;* this is indicated by the numbers of circles in the two areas. This example brings out the two meanings of "common" quite nicely. Which one you emphasize will doubtless depend on whether your interest lies in having empty auger holes or many bees to pollinate your lucerne. The wood where Kluijver put additional nesting boxes for *Parus* was like *A*, except that some were not used; and the other more natural wood, which he described as lacking a sufficiency of tree-holes for nests, was like *B* (sec. 12.31).

The next example is more complex because it deals with living food instead of a lifeless nonexpendable resource like nesting sites. Suppose that, in Figure 14.07, *A* and *B* are two areas where prickly pears grow in colonies; the two areas are alike in climate, soil, distribution of suitable places for *Opuntia* to grow, and every other respect except that *A* supports a population of *Cactoblastis cactorum*, whereas in *B* there is a variant which has very much poorer powers of dispersal. Say we call it *Cactoblastis blastorum*. Area *C* is like the

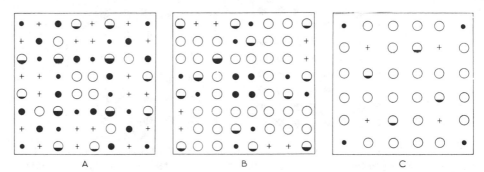

Fig. 14.07.—The populations of the large areas *A*, *B*, and *C* are made up of a number of local populations. Not all the favorable localities support local populations (*hollow circles*). Favorable localities are distributed equally densely in *A* and *B* but more sparsely in *C*. Area *A* is occupied by a species with high powers of dispersal; *B* and *C* are occupied by a species which is similar except that it has inferior powers of dispersal. The symbols are taken from Fig. 14.02. In Area *A* the number of animals is chiefly limited by the stock of food in the absolute sense. In *B* and *C* there is no absolute shortage of food, but the numbers of animals in these areas are limited by a shortage of food in the relative sense (sec. 11.12). This is related to the poor powers of dispersal of the species that lives in these areas. For further explanation see text.

other two except that, for physiographical reasons, colonies of prickly pears must remain very sparsely distributed, irrespective of whether any are destroyed by *Cactoblastis* or not. Area *C* supports a population of *C. blastorum*. The distributions of the circles and the shading in them indicate: in area *A* there are few colonies of prickly pears, but *C. cactorum* is making good use of what are there. Its numbers are clearly being limited chiefly by the absolute amount of food in the area; and the numerous crosses indicate that there is definitely less food in the area as a result of the presence of *C. cactorum*. In area *B* there are more colonies of prickly pears. Relatively few of them harbor local populations of *C. blastorum*. The numbers of *C. blastorum* are not being seriously limited by an absolute shortage of food. Nevertheless, by virtue of their poor powers of dispersal, they are suffering from a relative shortage of food. In area *C* the circumstances are much the same as in *B*, only more so. The food is even harder to find; the death-rate is therefore higher still; and the numbers of *C. blastorum* in the area are few indeed. The presence of a few crosses in both *B* and *C* shows that relative shortage of food (sec. 11.12) is the real cause of the trouble, because, once *C. blastorum* has found a colony of prickly pears, the insects increase rapidly and, in due course, destroy all the food, just as the real *C. cactorum* does in area *A*.

We need not have invented *C. blastorum* for area *B*. Several natural examples have been described, for example, *Chrysomela gemellata* (sec. 11.12) and *Rhizobius ventralis* (sec. 10.321). The fact that the latter feeds on animals instead of plants does not alter the principle. We do not know of an exact parallel for area *C*, but *Thrips imaginis* near Adelaide during summer are kept

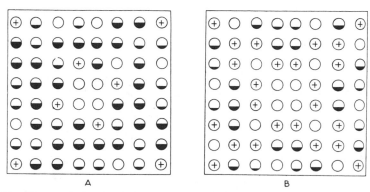

A   B

Fig. 14.08.—The populations in the large areas *A* and *B* are made up of a number of local populations. There is an equal number of suitable places where the animals may live in each area. In area *B* there is an active predator whose powers of dispersal and multiplication match those of the prey. This predator is absent from *A*, where its place is taken by one which is more sluggish. Relatively more of the suitable places are occupied by the prey in *A* than in *B*, and the local populations are, on the average, larger. There are more local situations in *B* from which the prey has recently been exterminated. Altogether, the animal is more common in *A* than in *B*. The symbols have the same meaning as in Fig. 14.03.

scarce by a relative shortage of food as *C. blastorum* is in area *C*, but they do not destroy their food so thoroughly as *C. blastorum* is supposed to do.

Let us consider next the case of an animal which has an active predator in its environment. We take the symbols from Figure 14.03 to make Figure 14.08. The two areas, *A* and *B*, have the same number of circles, because the presence of the animals does not influence the stocks of food, etc., in the area. Both *A* and *B* support populations of the same sort of herbivores. But in *B* there is also a species which is an active predator, with powers of dispersal and rate of increase which match those of the prey; in area *A* these predators are replaced by another species with equivalent capacity for increase but inferior dispersive powers. In *B* many localities which are quite favorable are empty; some have become empty only recently (circle with cross in it), and in most of those which are occupied the numbers are low. Either the herbivores have only recently arrived at the place and have not yet had time to become numerous, or else, if they have been there longer, the predators have found them and are in the process of exterminating the local colony. In *A*, relatively more of the favorable places are occupied, and the numbers in the local populations are, on the whole, larger. There are relatively few places from which the prey have been recently exterminated. All this can be explained simply by saying that the prey, being superior to the predators in dispersive powers, enjoy, on the average, a relatively long period of freedom from attack in each new place that they colonize. Several natural examples of this principle were discussed in section 10.321.

A similar diagram may be drawn to illustrate the case in which the numbers

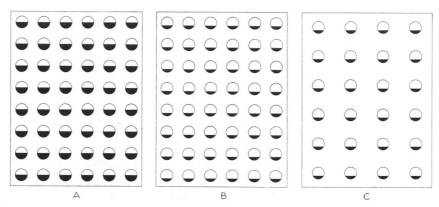

A   B   C

FIG. 14.09.—The populations in the large areas *A*, *B*, and *C* are made up of a number of local populations. The climate in *A* is more favorable than that in *B* or *C*; so the numbers (indicated by the shaded part of the circle) are, on the average, larger in each locality in *A* and in the area as a whole. There are fewer favorable localities in *C* than in *B*; so there are, on the average, fewer animals in area *C* than in *B*, where the climate is the same but the number of places greater.

in the area are determined largely by climate. In Figure 14.05 the two areas *A* and *B* were considered to differ only with respect to climate. The circles in Figure 14.09 may be related to Figure 14.05. In Figure 14.09 there is the same number of suitable places where the animals may live in *A* as in *B*, but the climate is more favorable in *A* than in *B*, so the animals are more common relative to their stocks of food, etc., in *A* than in *B*; there are also more of them in the area. Area *C* is inferior to *B* with respect to the number of places, and it is inferior to *A* with respect to both the number of places and the climate. The animals are uncommon in *C* relative to *A* in both meanings of the word. The natural population of *Porosagrotis orthogonia* (sec. 13.13) is a particular example of this principle, except that the difference between *A* and *B* in that case referred to the difference between the same area before and after it had been developed for agriculture.

The reader may have anticipated the general conclusion that we have been leading up to. The numbers of animals in a natural population may be limited in three ways: (*a*) by shortage of material resources, such as food, places in which to make nests, etc.; (*b*) by inaccessibility of these material resources relative to the animals' capacities for dispersal and searching; and (*c*) by shortage of time when the rate of increase *r* is positive. Of these three ways, the first is probably the least, and the last is probably the most, important in nature. Concerning *c*, the fluctuations in the value of *r* may be caused by weather, predators, or any other component of environment which influences the rate of increase. For example, the fluctuations in the value of *r* which are determined by weather may be rhythmical in response to the progression of the seasons (e.g., *Thrips imaginis*, sec. 13.11) or more erratic in response to "runs" of years with "good" or "bad" weather (e.g., *Austroicetes cruciata*, sec. 13.12). The fluctuations in *r* which are determined by the activities of predators must be considered in relation to the populations in local situations (Figs. 14.03 and 14.08). How long each newly founded colony may be allowed to multiply free from predators may depend on the dispersive powers of the predators relative to those of the prey.

With respect to the second way in which numbers may be limited, the food or other resource may be inaccessible either because it is sparsely distributed, like the food of *Thrips imaginis* during late summer (sec. 13.11), or because it is concealed, like the food of *Lygocerus* (sec. 10.321). In either case the inaccessibility is relative to the animals' capacities for dispersal and searching. We emphasize, in section 11.12 and elsewhere, that it is the inaccessibility of the resource which is important; the argument is independent of whether the stocks of food, etc., are scarce or plentiful, in the absolute sense of so much per square mile.

Of course, it is not to be supposed that the ecology of many natural populations would be so simple that their numbers would be explained neatly by any one of the principles described in this and the preceding two sections. A large number of systems of varying complexity could be synthesized, in the imagination, from various combinations of the principles set out in these sections. But this is not their purpose: they are intended to help the student to analyze the complex systems which he finds in nature into their simpler components, so that he may understand them better.

14.3 SOME PRACTICAL CONSIDERATIONS ABOUT THE NUMBERS OF ANIMALS IN NATURAL POPULATIONS

Darwin in chapter 3 of *The Origin of Species* wrote: "The causes which check the natural tendency of each species to increase in numbers are most obscure. Look at the most vigorous species; by as much as it swarms in numbers, by so much will its tendency to increase be still further increased. We know not what the checks are in one single instance. Nor will this surprise any one who reflects on how ignorant we are on this head." A little further on he wrote: "It is good thus to try in our imagination to give any form some advantage over another. Probably in no single instance should we know what to do, so as to succeed." In chapter 11 he stated: "Rarity is the attribute of a vast number of species of all classes in all countries. If we ask ourselves why this or that species is rare, we can answer that something is unfavourable in its conditions of life; but what that something is we can hardly ever tell."

During the century that has passed since these passages were written, we have learned enough about "the conditions of life" of at least a few species to be able, at will, to alter their numbers greatly. And we have, unwittingly, altered the "conditions of life" for a great many more species, so that their numbers have increased, or decreased, sometimes to our advantage, sometimes to our disadvantage.

With species that are pests of crops, foodstuffs, clothing, etc., it is to our advantage if we can insure that they never become numerous enough to eat a considerable proportion of their total resources of food. The majority of such species are insects. With respect to most of them, our ignorance is still abysmal, and the only way we know how to alter their "conditions of life" to our advantage is to poison as many of them as we can with insecticides. The chief disadvantage of the method is its costliness; but as the insecticides increase in number and complexity, we run an increasing risk of poisoning ourselves as well as the insects.

Ecologically, the use of insecticides fits into the principle illustrated by Figure 14.03. The difference between a good insecticide and a poor one is illustrated, in Figure 14.05, by the middle pair of curves. The difference between an insecticide applied frequently and one used infrequently is shown by the top pair of curves in Figure 14.05. And the difference between an insecticide applied thoroughly and one applied less well is shown in Figure 14.09. It is incorrect to say that, because an insecticide is not a "density-dependent factor," it cannot "regulate" the numbers in a population. There are millions of farmers who could testify that the proper use of insecticides makes a big difference to the size of the populations of certain insects. Many people all over the world would starve if this were not so.

"Biological control" (that is, the introduction of new sorts of predators into the area) is the next most popular way of altering the "conditions of life" of an insect pest. But other methods may be suggested by the principles set out in Figures 14.05 and 14.09. Modifications in husbandry might be thought out which would shorten the period available to the insect for multiplication or reduce the value of r (Fig. 14.05). Or other methods might be tried which would reduce the number of suitable places for the insects to live in or alter the distribution of such places in a way that would be detrimental to the pest (Fig. 14.09).

With animals that are valued for their flesh or fur or for some other quality, it is necessary to take quite a different point of view. It matters hardly at all whether the populations use up much or little of their total resources of food, etc., so long as the individuals are numerous enough to be hunted or fished with profit. If it so happened that the population conformed to Figure 14.01 or Figure 14.02, then the numbers per square mile might be increased by adding to the stocks of the limiting resource in the area. This method seems to have been greatly successful with the muskrat in North America (sec. 12.31). But if it happened that the population conformed to the principles illustrated by Figures 14.03–14.05, then other methods would be required.

14.4 SOME REFLECTIONS ON THE GREATLY MISUNDERSTOOD SUBJECT OF "EXTINCTION"

We conclude this chapter with an anticlimax. The reader may regard it as an appendix if he likes. However rare they may be with respect to the proportion of their total resources that they use up, most of the species that are studied or observed are common in terms of the number of individuals per square mile. This is easily explained by a number of practical reasons. But the fact that we usually study unusual populations is far too frequently overlooked. The truth is that the vast majority of species are rare, by whatever criterion (sec. 14.1) they are judged. Smith (1935, p. 880) commented on this: "The fact that the number of species which become sufficiently abundant to damage crops is *relatively* small, and that such species form only an insignificant fraction of the total number of phytophagous insects is ignored." Bodenheimer (1930, in Smith, 1935) commented in the same vein: "It is only in rare borderline cases that the food is used up to the possible limit. Any meadow, field, or orchard will prove this sufficiently." And, of course, Darwin was well aware of the fact. In a slightly different context he wrote (chap. 11 of *The Origin of Species*): "To admit that species generally become rare before they become extinct, to feel no surprise at the rarity of the species, and yet to marvel greatly when the species ceases to exist, is much the same as to admit that sickness in the individual is the forerunner of death—to feel no surprise at sickness, but, when the sick man dies, to wonder and to suspect that he died of some deed of violence."

Nevertheless, the misconception prevails that the extinction of a population is a very rare event. This leads our colleagues who hold to the dogma of "density-dependent factors" to propound this riddle. On hearing us expound our views on ecology, they ask: "How is it, if there is no density-dependent factor in the environment, that the population does not become extinct?" This places us in a position like that of the man in the dock who was asked to answer Yes or No to the question: "Do you still beat your wife?" We cannot answer the question until we have cleared up the misconception in the mind of the questioner.

First of all, it is indeed true that the species which we study are less likely than most to become extinct, because invariably we choose for study those in which the populations are large. Elsewhere in this book, and especially in sections 13.11 and 13.12, we explain why certain species of insects did not become extinct during the period that they were studied and infer that they are not likely to become extinct during the immediate future. A quotation from Stevenson-Hamilton (1937, p. 258) shows that the same principles apply to mammals which commonly maintain large populations: "The extermination, under natural conditions, of one indigenous species by another, whether directly through carnivores consuming herbivorous types, or indirectly by one herbivorous type proving too strong for its associates in the same area, except as a final culmination, after Man, or one of the other factors cited above, has first played the principal part, is unknown in natural history so far as our experience extends, and may be safely ruled out in any wild-life reservation *of adequate extent where room exists for seasonal migration* [our italics]."

But the risk of extinction for species in which the populations are small is greater. Darwin in the fourth chapter of *The Origin of Species* wrote: "Any form which is represented by a few individuals will run a good chance of utter extinction during great fluctuations in the nature of the seasons, or from a temporary increase in the numbers of its enemies." A quotation from Ford (1945a, p. 143) shows how this risk operated in the particular case of the butterfly known as the "Wood White":

Any species can survive such periodic fluctuation provided that its numbers are large enough, and that it is somewhere sufficiently well established to tide over the dangerous period when it is reduced to its lowest level. Clearly, a "normal" cycle of this kind may be disastrous to a butterfly which maintains itself precariously, whether in an isolated locality or in the country as a whole. The Wood White survived in Westmorland until about 1905 when it disappeared and has never been seen there again. At that time the species was in general becoming rare, and in the south retracting its range to a few favoured places. From these it could, and did, spread once more; but not in the north, where the one isolated colony was wiped out by a process which had no serious consequences elsewhere. Similarly, it disappeared from the New Forest early this century and has not returned. It was said at the time that this was due to overcollecting, and probably that was true. But I suspect that the butterfly was reduced to dangerously small numbers by natural causes, operating there as elsewhere, and for that reason the activities of the collector were fatal.

Ford quoted several other examples of butterflies which have recently become extinct in Britain.

There is evidence from paleontology to show that extinction of species is commonplace in the time scale of that science. Simpson (1952) estimated that there might now be 2,000,000 species of plants and animals in the world and that the total number of species that have existed since the "dawn of life" may be of the order of 500,000,000. On this estimate, more than 99 per cent of the species that have ever existed are now extinct. Mayr (1942, p. 224) pointed out that many of the species that are known to have become extinct during modern times have lived on small islands, where the terrain would be more uniform and the opportunity for dispersal less than in a larger area.

There is no fundamental distinction to be made between the extinction of a local population and the extinction of a species other than this that the species becomes extinct with the extinction of the last local population. We can witness the extinction of local populations, of even the most abundant species, going on all around us all the time. So if the extinction of a population were proof that the "environment" lacked a "density-dependent factor," we would have ample evidence of the absence of "density-dependent factors" from the "environments" of the animals in local populations of most species.

Species are likely to be "rare" near the margins of their distributions, and outside the distribution they are "extinct." Because distribution and abundance are but two aspects of one phenomenon, the study of abundance in different parts of the distribution is itself a study of the causes of rareness and commonness in species.

Bibliography and Author Index

The page citations following the items in the Bibliography give the location of reference to the given title in the text and replace the customary author index.

AHMAD, T. 1936. The influence of ecological factors on the Mediterranean flour moth *Ephestia kühniella* and its parasite *Nemeritus canescens,* J. Anim. Ecol., 5 : 67–93. Pp. 54, 55, 56.

ALLAN, P. F. 1939. Development of ponds for wildlife in the southern high plains, Tr. North Am. Wildlife Conf., 4 : 339–42. P. 133.

ALLEE, W. C.; EMERSON, A. E.; PARK, O.; PARK, T.; and SCHMIDT, K. P. 1949. Principles of animal ecology. Philadelphia: W. B. Saunders Co. Pp. 25, 26, 114, 156.

ANDREWARTHA, H. G. 1935. Thrips investigation No. 7. On the effect of temperature and food upon egg production and the length of adult life of *Thrips imaginis,* Bagnall., J. Counc. Scient. & Indust. Research Australia, 8 : 281–88. Pp. 68, 70, 157.

———. 1937. Locusts and grasshoppers in South Australia: some records of past outbreaks, J. Dept. Agr. South Australia, 41 : 366–68. P. 135.

———. 1939. The small plague grasshopper (*Austroicetes cruciata,* Sauss.), *ibid.,* 43 : 99–106. Pp. 121, 171.

———. 1940. The environment of the Australian plague locust (*Chortoicetes terminifera,* Walk.) in South Australia, Tr. Roy Soc. South Australia, 64 : 76–94. Pp. 136, 140, 141, 142.

———. 1943*a*. The significance of grasshoppers in some aspects of soil conservation in South Australia and Western Australia, J. Dept. Agr. South Australia, 46 : 314–22. Pp. 171, 172.

———. 1943*b*. Diapause in the eggs of *Austroicetes cruciata* Sauss. (Acrididae) with particular reference to the influence of temperature on the elimination of diapause, Bull. Ent. Research, 34 : 1–17, P. 171.

———. 1944*a*. Air temperature records as a guide to the date of hatching of the nymphs of *Austroicetes cruciata,* Sauss. (Orthoptera), *ibid.,* 35 : 31–41. Pp. 62, 64, 65, 66, 171.

———. 1944*b*. The distribution of plagues of *Austroicetes cruciata,* Sauss. (Acrididae) in Australia in relation to climate, vegetation and soil, Tr. Roy. Soc. South Australia, 68 : 315–26. Pp. 171, 172, 175, 177, 178, 182.

———. 1944*c*. The influence of temperature on the elimination of diapause from the eggs of the race of *Austriocetes cruciata* Sauss. (Acrididae) occurring in Western Australia, Australian J. Exper. Biol. & M. Sc., 22 : 17–20. P. 171.

———. 1970. Introduction to the study of animal populations. 2nd ed. London: Chapman & Hall, Ltd. P. x.

ANDREWARTHA, H. G. and BIRCH, L. C. 1948. Measurement of "environmental resistance" in the Australian plague grasshopper, Nature, 161 : 447–48. P. 171.

———. 1953. The Lotka-Volterra theory of interspecific competition, Australian J. Zoöl., 1 : 174–77. P. 407.

ANDREWARTHA, H. G.; DAVIDSON, J.; and SWAN, D. C. 1938. Vegetation types associated with plague grasshoppers in South Australia, Bull. Dept. Agr. South Australia, No. 333. P. 171.

ANDREWARTHA, H. V. 1936. The influence of temperature on the rate of development of the immature stages of *Thrips imaginis* Bagnall and *Haplothrips victoriensis* Bagnall, J. Counc. Scient. & Indust. Research Australia, 9 : 57–64. P. 157.

BACHMETJEW, P. 1907. Experimentelle entomologische Studien vom physikalisch-chemischen Standpunkt aus. Vol. 2 : Einfluss der äussern Factoren auf Insekten. Leipzig: Engelmann. P. 86.

BARBER, G. W., and DICKE, F. F. 1939. Effect of temperature and moisture on overwintering pupae of the corn earworm in the north-eastern states, J. Agr. Research, 59 : 711–23. Pp. 90, 92.

BEALL, G. 1940. The fit and significance of contagious distributions when applied to observations on larval insects, Ecology, 21 : 460–74. P. 150.

BÉLEHRÁDEK, J. 1935. Temperature and living matter. Berlin: Borntraeger. Pp. 42, 43, 44.

BIRCH, L. C. 1942. The influence of temperatures above the developmental zero on the development of the eggs of *Austroicetes cruciata* Sauss. (Orthoptera), Australian J. Exper. Biol. & M. Sc., 20 : 17–25. Pp. 26, 58, 171.

————. 1944*a*. An improved method for determining the influence of temperature on the rate of development of insect eggs using eggs of the small strain of *Calandra oryzae* L. (Coleoptera), *ibid.*, 22 : 277–83. Pp. 47, 48, 49.

————. 1945*a*. The mortality of the immature stages of *Calandra oryzae* L. (small strain) and *Rhizopertha dominica* Fab. in wheat of different moisture contents, *ibid.*, 23 : 141–45. P. 14.

————. 1945*c*. The influence of temperature on the development of the different stages of *Calandra oryzae* L. and *Rhizopertha dominica* Fab. (Coleoptera), Australian J. Exper. Biol. & M. Sc., 23 : 29–35. Pp. 47, 57, 67.

————. 1945*d*. The influence of temperature, humidity and density on the oviposition of the small strain of *Calandra oryzae* L. and *Rhizopertha dominica* Fab. (Coleoptera), *ibid.*, pp. 197–203. P. 67.

————. 1946*a*. The heating of wheat stored in bulk in Australia, J. Australian Inst. Agr. Sc., 12 : 27–31. Pp. 15, 23.

————. 1946*b*. The movements of *Calandra oryzae* L. (small strain) in experimental bulks of wheat, *ibid.*, pp. 21–26. P. 33.

————. 1947. The ability of flour beetles to breed in wheat, Ecology, 28 : 322–24. P. 47.

————. 1948. The intrinsic rate of natural increase of an insect population, J. Anim. Ecol., 17 : 15–26. Pp. 4, 6, 7, 13, 14, 15.

————. 1953*a*. Experimental background to the study of the distribution and abundance of insects. I. The influence of temperature, moisture and food on the innate capacity for increase of three grain beetles, Ecology, 34 : 698–711. Pp. 4, 9, 17, 18, 19, 22, 23.

————. 1953*b*. Experimental background to the study of the distribution and abundance of insects. II. The relation between innate capacity for increase in numbers and the abundance of three grain beetles in experimental populations, *ibid.*, 34 : 712–26. P. 4.

BIRCH, L. C., and ANDREWARTHA, H. G. 1941. The influence of weather on grasshopper plagues in South Australia, J. Dept. Agr. South Australia, 45 : 95–100. Pp. 171, 178.

————. 1942. The influence of moisture on the eggs of *Austroicetes cruciata* Sauss. (Orthoptera) with reference to their ability to survive desiccation, Australian J. Exper. Biol. & M. Sc., 20 : 1–8. P. 171.

————. 1944. The influence of drought on the survival of eggs of *Austroicetes cruciata* Sauss. (Orthoptera) in South Australia, Bull. Ent. Research, 35 : 243–50. Pp. 171, 176, 177.

BIRCH, L. C., and SNOWBALL, G. J. 1945. The development of the eggs of *Rhizopertha dominica* Fab. (Coleoptera) at constant temperatures, Australian J. Exper. Biol. & M. Sc., 23 : 37–40. P. 47.

BLISS, C. I. 1926. Temperature characteristics for prepupal development in *Drosophila melanogaster*, J. Gen. Physiol., 9 : 467–95. P. 43.

BLISS, C. I., and FISHER, R. A. 1953. Fitting the negative binomial distribution to biological data and note on the efficient fitting of the negative binomial, Biometrics, 9 : 176–200. P. 156.

BODENHEIMER, F. S. 1930. Über die Grundlagen einer allgemeinen Epidemiologie der Insektenkalamitäten, Ztschr. f. Angew. Ent., 16 : 433–50. P. 251.

BOGERT, C. M. 1952. Relative abundance, habitats and normal thermal levels of some Virginian salamanders, Ecology, 33 : 16–30. P. 32.

BRETT, J. R. 1944. Some lethal temperature relations of Algonquin Park fishes, Univ. Toronto Stud. Biol., Vol. 52. Pp. 97, 98.

BROWNING, T. O. 1952*b*. The influence of temperature on the rate of development of insects, with special reference to the eggs of *Gryllulus commodus* Walker, *ibid.*, pp. 96–111. Pp. 47, 48, 49, 50, 51.

BRUES, C. T. 1939. Studies on the fauna of some thermal springs in the Dutch East Indies, Proc. Am. Acad. Arts & Sc., 73 : 71–95. P. 26.

BUXTON, P. A. 1924. Heat, moisture, and animal life in deserts, Proc. Roy. Soc. London, B, 96 : 123–31. P. 100.

————. 1932*b*. The climate in which the rat-flea lives, Indian J. M. Research, 20 : 281–97. P. 120.

CHAPMAN, R. N.; MICKEL, C. E.; PARKER, J. R.; MILLER, G. E.; and KELLY, E. G. 1926. Studies in the ecology of sand dune insects, Ecology, 7 : 416–26. P. 100.

CLARK, L. R. 1947*a*. An ecological study of the Australian plague locust *Chortoicetes terminifera* Walk. in the Bogan-Macquarie outbreak area in N.S.W., Bull. Counc. Scient. & Indust Research Australia, No. 226. Pp. 30, 31, 136, 137, 138, 139, 142, 146.

————. 1949. Behaviour of swarm hoppers of the Australian plague locust, *Chortoicetes terminifera* Walk., *ibid.*, No. 245. P. 31.

COLE, L. C. 1946a. A study of the Cryptozoa of an Illinois woodland, Ecol. Monogr., 16 : 49–86. P. 150.

————. 1946*b*. A theory for analyzing contagiously distributed populations, Ecology, 27 : 329–41. P. 150.

————. 1951. Population cycles and random oscillations, J. Wildlife Management, 15 : 233–52. Pp. 231, 232, 233, 234.

————. 1954. Some features of random population cycles, *ibid.*, 18 : 2–24. P. 233.

COOK, W. C. 1924. The distribution of the pale western cutworm (*Porosagrotis orthogonia* Morr.): a study in physical ecology, Ecology, 5 : 60–69. P. 182.

————. 1926. Some weather relations of the pale western cutworm (*Porosagrotis orthogonia* Morr.). A preliminary study, *ibid.*, 7 : 37–47. P. 182.

————. 1930. Field studies of the pale western cutworm (*Porosagrotis orthagonia* Morr.), Bull. Montana Agr. Exper. Sta., No. 225. Pp. 182, 183, 185, 186, 187.

COWLES, R. B. 1945. Heat-induced sterility and its possible bearing on evolution, Am. Nat., 79 : 160–75. P. 71.

CRABB, W. D. 1948. The ecology and management of the prairie spotted skunk in Iowa, Ecol. Monogr., 18 : 201–32. P. 111.

DARBY, H. H., and KAPP, E. M. 1933. Observations on the thermal death points of *Anastrepha ludens* (Loew.), Tech. Bull. U.S. Dept. Agr. No., 400. Pp. 98, 99.

DAVIDSON, J. 1931. The influence of temperature on the incubation period of the eggs of *Sminthurus viridis* L. (Collembola), Australian J. Exper. Biol. & M. Sc., 9 : 143–52. P. 26.

————. 1936a. On the ecology of the black-tipped locust (*Chortoicetes terminifera,* Walk.) in South Australia, Tr. Roy. Soc. South Australia, 60 : 137–52. P. 135.

————. 1936b. The apple-thrips (*Thrips imaginis* Bagnall) in South Australia, J. Dept. Agr. South Australia, 39 : 930–39. P. 157.

————. 1936c. Climate in relation to insect ecology in Australia. 3. Bioclimatic zones in Australia, Tr. Roy. Soc. South Australia, 60 : 88–92. Pp. 158, 165, 181.

————. 1942. On the speed of development of insect eggs at constant temperatures, Australian J. Exper. Biol. & M. Sc., 20 : 233–39. Pp. 45, 47.

————. 1943a. The time required for the eggs of the body louse (*Pediculus humanus corporis* de Geer) to develop and hatch at different temperatures, M. J. Australia, June 12, 1943, pp. 533–36. P. 47.

————. 1943b. On the speed of development of insect eggs at constant temperatures, Australian J. Exper. Biol. & M. Sc., 20 : 233–39. P. 47.

————. 1944. On the relationship between temperature and rate of development of insects at constant temperatures, J. Anim. Ecol., 13 : 26–38. Pp. 41, 42, 45, 46, 47, 48.

DAVIDSON, J., and ANDREWARTHA, H. G. 1948a. Annual trends in a natural population of *Thrips imaginis* (Thysanoptera), J. Anim. Ecol., 17 : 193–99. Pp. 37, 38, 157, 160, 163.

————. 1948b. The influence of rainfall, evaporation and atmospheric temperature on fluctuations in the size of a natural population of *Thrips imaginis* (Thysanoptera), *ibid.,* pp. 200–222. Pp. 37, 39, 157, 161, 162, 163, 165, 167.

DAVIS, D. E. 1953. The characteristics of rat populations, Quart. Rev. Biol., 28 : 373–401. P. 112.

DEAL, J. 1941. The temperature preferendum of certain insects, J. Anim. Ecol., 10 : 323–55. P. 29.

DEEVEY, E. S., JR. 1947. Life tables for natural populations of animals, Quart. Rev. Biol., 22 : 283-314. P. 9.

DICK, J. 1937. Oviposition in certain Coleoptera, Ann. Appl. Biol., 24 : 762–96. Pp. 67, 69, 71.

DITMAN, L. P.; VOGT, G. B.; and SMITH, D. R. 1943. The relation of unfreezable water to cold-hardiness of insects, J. Econ. Ent., 36 : 304–11. Pp. 88, 90.

DOBZHANSKY, TH., and PAVAN, C. 1950. Local and seasonal variations in relative frequencies of species of *Drosophila* in Brazil, J. Anim. Ecol., 19 : 1–14. P. 153.

DOBZHANSKY, TH., and WRIGHT, S. 1943. Genetics of natural populations. X. Dispersion rates in *Drosophila pseudoobscura*, Genetics, 28 : 304–40. Pp. 35, 36, 37.

DOUDOROFF, P. 1938. Reactions of marine fishes to temperature gradients, Biol. Bull., 75 : 494–509. Pp. 27, 28, 29.

DOWDESWELL, W. H.; FISHER, R. A.; and FORD, E. B. 1940. The quantitative study of populations in the Lepidoptera. I. *Polyommatus icarus*, Ann. Eugenics, 10 : 123–36. P. 217.

DUBLIN, L. I., and LOTKA, A. J. 1925. On the true rate of increase as exemplified by the population of the United States, 1920, J. Am. Statist. A., 20 : 305–39. Pp. 11, 13.

DUBLIN, L. I.; LOTKA, A. J.; and SPIEGELMAN, M. 1949. Length of life. Rev. ed. New York: Ronald Press Co. P. 10.

DUVAL, M., and PORTIER, P. 1922. Limite de résistance au froid des chenilles de *Cossus cossus*, C.R. Soc. biol., Paris, 86 : 2–4. P. 85.

DYMOND, J. R. 1947. Fluctuations in animal populations with special reference to those of Canada, Tr. Roy. Soc. Canada, 41 : 1–34. Pp. 133, 229, 230, 231, 232.

ELTON, C. 1927. Animal ecology. London: Sidgwick & Jackson. Pp. 104, 114.

———. 1938. Animal numbers and adaptation. *In:* BEER, R. G. DE (ed.), Evolution: essays on aspects of evolutionary biology, pp. 127–37. Oxford: Clarendon Press. P. 145.

———. 1949. Population interspersion: an essay on animal community patterns, J. Ecol., 37 : 1–23. Pp. 113, 168, 236.

ELTON, C., and NICHOLSON, M. 1942a. Fluctuations in numbers of the muskrat (*Ondatra zibethica*) in Canada, J. Anim. Ecol., 11 : 96–125. P. 229.

———. 1942b. The ten-year cycle in numbers of the lynx in Canada, *ibid.*, pp. 215–43. Pp. 228, 234.

ERRINGTON, P. L. 1934. Vulnerability of bobwhite populations to predation, Ecology, 15 : 110–27. Pp. 130, 131.

———. 1939. Reactions of muskrat populations to drought, *ibid.*, 20 : 168–86. P. 128.

———. 1943. An analysis of mink predation upon muskrats in north-central United States, Research Bull. Iowa Agr. Exper. Sta., 320 : 797–924. P. 128.

———. 1944. Ecology of the muskrat, Rep. Iowa Agr. Exper. Sta., 1944, pp. 187–89. Pp. 132, 133.

———. 1945. Some contributions of a fifteen-year local study of the northern bobwhite to a knowledge of population phenomena, Ecol. Monogr., 15 : 1–34. Pp. 146, 224, 225.

———. 1946. Predation and vertebrate populations, Quart. Rev. Biol., 21 : 145–77 and 221–45. Pp. 130, 133.

———. 1948. Environmental control for increasing muskrat production, Tr. 13th North Am. Wildlife Conf., pp. 596–609. Pp. 132, 133.

ERRINGTON, P. L., and HAMERSTROM, F. N. 1936. The northern bobwhite's winter territory, Research Bull. Iowa Agr. Exper. Sta., Vol. 201. Pp. 130, 131.

ERRINGTON, P. L., and SCOTT, T. G. 1945. Reduction in productivity of muskrat pelts on an Iowa marsh through depredations of red foxes, J. Agr. Research, 71 : 137–48. P. 128.

EVANS, F. C. 1949. A population study of house mice (*Mus musculus*) following a period of local abundance, J. Mammal, 30 : 351–63. P. 218.

EVANS, F. C., and SMITH, F. E. 1952. The intrinsic rate of natural increase for the human louse *Pediculus humanis* L., Am. Nat. 86 : 299–310. P. 4.

FINNEY, D. J. 1947. Probit analysis: a statistical treatment of the sigmoid response curve. Cambridge: At the University Press. Pp. 74, 75, 76.

FISHER, R. A. 1948. Statistical methods for research workers. 10th ed. Edinburgh: Oliver & Boyd. P. 166.

FISHER, R. A., and FORD, E. B. 1947. The spread of a gene in natural conditions in a colony of the moth *Panaxia dominula* (L.), Heredity, 1 : 143–74. P. 109.

FISHER, R. A., and YATES, F. 1948. Statistical tables for biological, agricultural and medical research. London: Oliver & Boyd. P. 76.

FLANDERS, S. E. 1949. Black scale, California Agriculture, 3 : 74. P. 129.

FORD, E. B. 1945a. Butterflies. London: Collins. Pp. 103, 154, 155, 218, 252.

FRAENKEL, G., and GUNN, D. L. 1940. The orientation of animals: kineses, taxes and compass reactions. Oxford: Clarendon Press. P. 29.

FRY, F. E. J. 1947. Effects of the environment on animal activity, Univ. Toronto Stud. Biol., 55 : 1–62. Pp. 28, 42, 43, 95, 96.

FRY, F. E. J.; BRETT, J. R.; and CLAUSEN, G. H. 1942. Lethal limits of temperature for young goldfish, Rev. canad. de biol., 1 : 50–56. P. 96.

FRY, F. E. J., and HART, J. S. 1948. Cruising speed of goldfish in relation to water temperature, J. Fish. Research Board Canada, 7 : 169–74. Pp. 33, 34.

FRY, F. E. J.; HART, J. S.; and WALKER, K. F. 1946. Lethal temperature relations for a
sample of young speckled trout *Salvelinus fontinalis,* Pub. Ontario Fish. Research Lab.,
66 : 5–35. Pp. 95, 97.

FULLER, M. E. 1934. The insect inhabitants of carrion: a study in animal ecology, Bull.
Counc. Scient. & Indust. Research Australia, No. 82. P. 115.

GASCHEN, H. 1945. Les Glossines de l'Afrique Occidentale Française, Acta trop., Suppl.,
Vol. 2. Pp. 217, 224.

GAY, F. J. 1953. Observations on the biology of *Lyctus brunneus* (Steph.), Australian J.
Zoöl., 1 : 102–10. P. 113.

GILMOUR, D.; WATERHOUSE, D. F.; and McINTYRE, G. A. 1946. An account of experiments
undertaken to determine the natural population density of the sheep blowfly, *Lucilia
cuprina* Wied., Bull. Counc. Scient. & Indust. Research Australia, No. 195. Pp. 150, 151,
152.

GIVEN, B. B. 1944. Notes on the physical ecology of *Diadromus (Thyraeella) collaris,* Grav.,
New Zealand J. Scient. Tech., 26 : 198–201. P. 83.

GLENN, P. A., 1922. Relation of temperature to development of the codlin moth, J. Econ.
Ent., 15 : 193–98. P. 62.

————. 1931. Use of temperature accumulations as an index to the time of appearance of
certain insect pests during the season, Tr. Illinois Acad. Sc., 24 : 167–80. P. 62.

GRANGE, W. B. 1947. Practical beaver and muskrat farming. Babcock, Wis.: Sandhill Press.
P. 133.

GUNN, D. L.; PERRY, F. C.; SEYMOUR, W. G.; TELFORD, T. M.; WRIGHT, E. M.; and
YEO, D. 1948. Behaviour of the desert locust (*Schistocerca gregaria,* Forsk.) in Kenya in
relation to aircraft spraying, Bull. Anti-locust Research Centre London, No. 3. P. 34.

GUNN, R. M. C.; SANDERS, R. N.; and GRANGER, W. 1942. Studies in fertility in sheep.
2. Seminal changes affecting fertility in rams, Bull. Counc. Scient. & Indust. Research,
Australia, No. 148. P. 71.

HARPER, J. L. 1977. Population biology of plants. London: Academic Press. P. x.

HARRIES, F. H. 1937. Some effects of temperature on the development and oviposition of
Microbracon hebetor (Say.), Ohio J. Sc., 37 : 165. P. 67.

————. 1939. Some temperature coefficients for insect oviposition, Ann. Ent. Soc. Amer.,
32 : 758–76. P. 68.

HART, J. S. 1952. Geographic variations of some physiological and morphological characters
in certain freshwater fish, Pub. Ontario Fish. Research Lab., Vol. 72. P. 98.

HENRY, C. J. 1939. Response of wildlife to management practices on the Lower Souris
Migratory Waterfowl Refuge, Tr. North Am. Wildlife Conf., 4 : 372–77. P. 133.

HENSON, W. R., and SHEPHERD, R. F. 1952. The effects of radiation on the habitat temperature of the lodgepole needle miner *Recurvaria milleri* (Gelechiidae, Lepidoptera), Canad.
J. Zoöl., 30 : 144–53. P. 119.

HEWITT, C. G. 1921. The conservation of the wild life of Canada. New York: Charles
Scribner's Sons. P. 229.

HODSON, A. C. 1937. Some aspects of the role of water in insect hibernation. Ecol. Monogr.,
7 : 271–315. P. 91.

HOLDAWAY, F. G. 1932. An experimental study of the growth of populations of the flour
beetle *Tribolium confusum* Duval as affected by atmospheric moisture, Ecol. Monogr.,
2 : 261–304. P. 3.

HOLME, N. A. 1950. Population dispersion in *Tellina tenuis,* da Costa, J. Marine Biol. A. United Kingdom, 29 : 267–80. P. 155.

HOWE, R. W. 1953a. Studies on beetles of the family Ptinidae. VIII. The intrinsic rate of increase of some ptinid beetles, Ann. Appl. Biol., 40 : 121–34. Pp. 4, 13, 15, 20, 22.

———. 1953b. The rapid determination of the intrinsic rate of increase of an insect population, *ibid.,* pp. 134–51. P. 4.

HOWE, R. W., and OXLEY, T. A. 1944. The use of carbon dioxide production as a measure of infestation of grain by insects, Bull. Ent. Research, 35 : 11–22. P. 15.

HUNTSMAN, A. G. 1946. Heat stroke in Canadian maritime stream fishes, J. Fish. Research Board Canada, 6 : 476–82. P. 97.

HUTCHINSON, G. E. 1953. The concept of pattern in ecology, Proc. Acad. Nat. Sc. Philadelphia, 105 : 1–12. P. 156.

IDE, F. P. 1935. The effect of temperature on the distribution of the mayfly fauna of a stream, Univ. Toronto Stud. Biol., 39 : 1–76. P. 59.

JACKSON, C. H. N. 1930. Contributions to the bionomics of *Glossina morsitans,* Bull. Ent. Research, 21 : 491–527. P. 217.

———. 1933. On the true density of tsetse flies, J. Anim. Ecol., 2 : 204–9. P. 217.

———. 1936. Some new methods in the study of *Glossina morsitans,* Proc. Zoöl. Soc. London, 1936, pp. 811–96. Pp. 220, 221.

———. 1939. The analysis of an animal population, J. Anim. Ecol., 8 : 238–46. P. 217.

———. 1949. The biology of tsetse flies, Biol. Rev., 24 : 174–97. P. 217.

JENKINS, D. W., and CARPENTER, S. J. 1946. Ecology of the tree hole breeding mosquitoes of Nearctic North America, Ecol. Monogr., 16 : 41–47. P. 113.

JOHNSON, C. G. 1940. The longevity of the fasting bed bug (*C. lectularius* L.) under experimental conditions and particularly in relation to the saturation deficiency law of water loss, Parasitology, 32 : 239–70. Pp. 75, 76, 77, 79, 80, 81.

KENDALL, M. G. 1948. Advanced theory of statistics. 4th ed. London: Charles Griffin & Co. Pp. 231, 234.

KENNEDY, J. S. 1939. The behaviour of the desert locust (*Schistocerca gregaria,* Forsk.) (Orthoptera) in an outbreak centre, Tr. Roy. Ent. Soc. London, 89 : 385–542. Pp. 32, 35.

KETTLE, D. S. 1951. The spatial distribution of *Culicoides impunctatus* Goet. under woodland and moorland conditions and its flight range through woodland, Bull. Ent. Research, 42 : 239–91. P. 126.

KEY, K. H. L. 1938. The regional and seasonal incidence of grasshopper plagues in Australia, Bull. Counc. Scient. & Indust. Research Australia, No. 117. P. 141.

———. 1942. An analysis of the outbreaks of the Australian plague locust (*Chortoicetes terminifera,* Walk.) during the seasons 1937–38 and 1938–39, *ibid.,* No. 146. P. 135.

———. 1943. The outbreak of the Australian plague locust (*Chortoicetes terminifera,* Walk.) in the season 1939–40, with special reference to the influence of climatic factors, *ibid.,* No. 160. P. 135.

———. 1945. The general ecological characteristics of the outbreak areas and outbreak years of the Australian plague locust (*Chortoicetes terminifera,* Walk.), *ibid.,* No. 186. Pp. 135, 136, 139, 140, 141.

———. 1954. The taxonomy, phases, and distribution of the genera *Chortoicetes* Brunn. and *Austroicetes* Uv. (Orthoptera: Acrididae). Canberra: Comm. Scient. & Indust. Research Organization. P. 136.

KLUIJVER, H. N. 1951. The population ecology of the great tit *Parus m. major,* Ardea, 39 : 1–135. Pp. 134, 135.

KNIGHT, H. H. 1922. Studies on the life history and biology of *Perillus bioculatus* including observations on the nature of the colour pattern, 19th Rep. State Ent. Minnesota, pp. 50–96. P. 94.

KOZHANCHIKOV, I. V. 1938. Physiological conditions of cold-hardiness in insects, Bull. Ent. Research, 29 : 252–62. Pp. 85, 87, 88, 89, 90.

KROGH, A. 1914. On the influence of the temperature on the rate of embryonic development, Ztschr. f. allg. Physiol., 16 : 163–77. P. 61.

LACK, D., and VENABLES, L. S. V. 1939. The habitat distribution of British woodland birds, J. Anim. Ecol., 8 : 39–70. Pp. 112, 114.

LARSEN, E. B. 1943. The influence of humidity on life and development of insects. Experiments on flies, Vidensk. Medd. dansk. naturh. Fören. Kbh., 107 : 127–84. P. 99.

LATHROP, F. H., and NICKELS, C. B. 1931. The blueberry maggot from an ecological viewpoint. Ann. Ent. Soc. Amer., 24 : 260–74. Pp. 200, 201.

LEES, A. D., and MILNE, A. 1951. The seasonal and diurnal activities of individual sheep ticks (*Ixodes ricinus* L.), Parasitology, 41 : 189–207. P. 213.

LEOPOLD, A. 1933. Game management. New York : Charles Scribner's Sons. Pp. 112, 130, 131.

———. 1943. Deer irruptions, Wisconsin Conserv. Bull., August, 1943. P. 1.

LESLIE, P. H., and PARK, T. 1949. The intrinsic rate of natural increase of *Tribolium castaneum* Herbst, Ecology, 30 : 469–77. P. 4.

LESLIE, P. H., and RANSON, R. M. 1940. The mortality, fertility and rate of natural increase of the vole (*Microtus agrestis*) as observed in the laboratory, J. Anim. Ecol., 9 : 27–52. Pp. 4, 6, 7, 8, 13, 14, 16.

LLOYD, L. 1943. Materials for a study in animal competition. II. The fauna of the sewage beds. III. The seasonal rhythm of *Psychoda alternata*, Say and an effect of intraspecific competition, Ann. Appl. Biol., 30 : 47–60, 358–64. Pp. 193, 194, 195, 197, 198.

LLOYD, L.; GRAHAM, J. F.; and REYNOLDSON, T. B. 1940. Materials for a study in animal competition. The fauna of the sewage bacteria beds, Ann. Appl. Biol., 27 : 122–50. P. 195.

LORENZ, K. Z. 1952. King Solomon's ring. London : Methuen Co. P. 103.

LOTKA, A. J. 1925. Elements of physical biology. Baltimore : Williams & Wilkins, Pp. 4, 12, 14.

———. 1939. Théorie analytique des associations biologiques. Deuxième partie. Analyse démographique avec application particulière à l'espèce humaine, "Actualités sc. indust.," 780 : 1–149. Paris: Hermann & Cie. P. 13.

LUDWIG, D. 1928. The effect of temperature on the development of an insect (*Popillia japonica*), Physiol. Zoöl., 1 : 358–98. P. 83.

LUDWIG, D., and CABLE, R. M. 1933. The effect of alternating temperatures on the pupal development of *Drosophila melanogaster*, Meigen, Physiol. Zoöl., 6 : 493–508. Pp. 56, 57, 58.

LUYET, B. J., and GEHENIO, M. P. 1940. Life and death at low temperatures. Normandy, Mo.: Biodynamica. ("Monographs on General Physiology," No. 1.) Pp. 73, 86.

McCABE, T. T., and BLANCHARD, B. D. 1950. Three species of *Peromyscus*. Santa Barbara, Calif.: Rood. P. 109.

MACAN, T. T., and WORTHINGTON, E. B. 1951. Life in lakes and rivers. London: Collins. P. 112.

McCLURE, H. E. 1943. Ecology and management of the mourning dove *Zenaidura macroura*, Research Bull. Iowa Agr. Exper. Sta., No. 310. P. 134.

McDONOGH, R. S. 1939. The habitat distribution and dispersal of the psychid moth *Luffia ferchaultella* in England and Wales, J. Anim. Ecol., 8 : 10–28. Pp. 105, 106, 107.

MACKENZIE, J. M. D. 1952. Fluctuations in the numbers of British tetraonids, J. Anim. Ecol., 21 : 128–53. Pp. 229, 230.

MACLEOD, J. 1934. *Ixodes ricinus* in relation to its physcial environment: the influence of climate on development, Parasitology, 26 : 282–305. P. 124.

MAIL, G. A. 1930. Winter soil temperatures and their relation to subterranean insect survival, J. Agr. Research, 41 : 571–92. Pp. 116, 117, 118.

———. 1932. Winter temperature gradients as a factor in insect survival, J. Econ. Ent., 25 : 1049–53. P. 116.

MAIL, G. A., and SALT, R. W. 1933. Temperature as a possible limiting factor in the northern spread of the Colorado potato beetle, J. Econ. Ent., 26 : 1068–75. Pp. 116, 117.

MAYR, E. 1942. Systematics and the origin of species. New York: Columbia University Press. P. 253.

MELLANBY, K. 1939. Low temperature and insect activity, Proc. Roy. Soc. London, B, 127 : 473–87. P. 73.

MILLER, J. M. 1931. High and low lethal temperatures for the western pine beetle, J. Agr. Research, 43 : 303–21. Pp. 93, 119.

MILNE, A. 1943. The comparison of sheep tick populations (*Ixodes ricinus* L.), Ann. Appl. Biol., 30 : 240–50. P. 209.

———. 1944. The ecology of the sheep tick, *Ixodes ricinus* L. Distribution of the tick in relation to soil and vegetation in northern England, Parasitology, 35 : 186–96. Pp. 125, 209.

———. 1945a. The ecology of the sheep tick, *Ixodes ricinus* L. The seasonal activity in Britain with particular reference to northern England, *ibid.*, 36 : 142–52. Pp. 209, 213.

———. 1945b. The ecology of the sheep tick, *Ixodes ricinus* L. Host availability and seasonal activity, *ibid.*, pp. 153–57. P. 209.

———. 1945c. The control of the sheep tick (*Ixodes ricinus* L.) by treatment of farm stock, Ann. Appl. Biol., 32 : 128–42. P. 209.

———. 1946. The ecology of the sheep tick, *Ixodes ricinus* L. Distribution of the tick on hill pasture, Parasitology, 37 : 75–81. Pp. 125, 209, 211, 212.

———. 1947a. The ecology of the sheep tick, *Ixodes ricinus* L. Some further aspects of activity, seasonal and diurnal, *ibid.*, 38 : 27–33. P. 209.

———. 1947b. The ecology of the sheep tick, *Ixodes ricinus* L. The infestations of hill sheep, *ibid.*, pp. 34–50. P. 209.

———. 1948. Pasture improvement and the control of sheep tick (*Ixodes ricinus* L.), Ann. Appl. Biol., 35 : 369–78. P. 209.

———. 1949. The ecology of the sheep tick, *Ixodes ricinus* L. Host relationships of the tick. 2. Observations on hill and moorland grazings in northern England, Parasitology, 39 : 173–97. Pp. 209, 214, 215.

———. 1950a. The ecology of the sheep tick, *Ixodes ricinus* L. Microhabitat economy of the adult tick, *ibid.*, 40 : 14–34. Pp. 124, 125, 126, 209.

———. 1950b. The ecology of the sheep tick, *Ixodes ricinus* (L.), Spatial distribution, *ibid.*, pp. 35–45. Pp. 124, 125, 209.

———. 1951. The seasonal and diurnal activities of individual sheep ticks (*Ixodes ricinus*, L.), *ibid.*, 41 : 189–208. P. 209.

———. 1952. Features of the ecology and control of the sheep tick, *Ixodes ricinus* L., in Britain, Ann. Appl. Biol., 39 : 144–46. Pp. 124, 209.

MOHR, C. D. 1943. Cattle droppings as ecological units, Ecol. Monogr., 13 : 275–98. P. 114.

MOORE, C. R., and QUICK, W. J. 1924. The scrotum as a temperature regulator for the testes, Am. J. Physiol., 68 : 70–79. P. 71.

MOORE, J. A. 1939. Temperature tolerance and rates of development in the eggs of Amphibia, Ecology, 20 : 459–78. P. 60.

———. 1940*a*. Adaptive differences in the egg-membranes of frogs, Am. Nat., 74 : 89–93. Pp. 26, 60.

———. 1940*b*. Stenothermy and eurothermy of animals in relation to habitat, *ibid.*, pp. 188–92. P. 27.

———. 1942. The rôle of temperature in the speciation of frogs, Biol. Symp., 6 : 189–213. P. 60.

MORAN, P. A. P. 1952. The statistical analysis of game-bird records, J. Anim. Ecol., 21 : 154–58. P. 231.

———. 1953. The statistical analysis of the Canadian lynx cycle. I. Structure and prediction. Australian J. Zoöl., 1 : 163–73. Pp. 228, 231, 234.

MUNGER, F. 1948. Reproduction and mortality of California red scales, resistant and non-resistant to hydrocyanic gas as affected by temperature, J. Agr. Research, 76 : 153–63. P. 83.

NAGEL, R. H., and SHEPARD, H. H. 1934. The lethal effect of low temperatures on the various stages of the confused flour beetle, J. Agr. Research, 48 : 1009–16. Pp. 80, 82.

NASH, T. A. M. 1930. A contribution to our knowledge of the bionomics of *Glossina morsitans*, Bull. Ent. Research, 21 : 201–56. P. 217.

———. 1933*a*. The ecology of *Glossina morsitans*, Westw., and two possible methods for its destruction. Part I, *ibid.*, 24 : 107–57. P. 217.

———. 1933*b*. The ecology of *Glossina morsitans*, Westw., and two possible methods for its destruction. Part II, *ibid.*, pp. 163–95. P. 217.

———. 1937. Climate, the vital factor in the ecology of *Glossina, ibid.*, 28 : 75–127. P. 217.

———. 1948. Tsetse flies in British West Africa. London: H. M. Stationery Office. Pp. 217, 219, 222, 223, 224.

NEL, R. G. 1936. The utilisation of low temperatures in the sterilisation of deciduous fruit infested with the immature stages of the Mediterranean fruit fly *Ceratitis capitata*, Scient. Bull. Dept. Agr. South Africa, No. 155. P. 82.

NICHOLSON, A. J. 1950. Population oscillations caused by competition for food, Nature, 165 : 476–77. P. 194.

NORRIS, M. J. 1933. Contributions towards the study of insect fertility. II. Experiments on the factors influencing fertility in *Ephestia kühniella* Z. (Lepidoptera, Phycitidae), Proc. Zoöl. Soc. London, 1933, pp. 903–34. P. 69.

ODUM, E. P. 1953. Fundamentals of ecology. Philadelphia, W. B. Saunders Co. P. 115.

PARKER, J. R. 1930. Some effects of temperature and moisture upon *Melanoplus mexicanus mexicanus* Saussure and *Camnula pellucida* Scudder (Orthoptera), Bull. Montana Agr. Exper. Sta., No. 223, Pp. 52, 53, 54.

PARKER, J. R.; STRAND, A. L.; and SEAMANS, H. L. 1921. Pale western cutworm (*Porosagrotis orthogonia*, Morr.), J. Agr. Research, 22 : 289–322. P. 182.

PARKIN, E. A. 1934. Observations on the biology of the *Lyctus* powder-post beetles, with special reference to oviposition and the egg, Ann. Appl. Biol., 21 : 495–518. P. 113.

PAYNE, N. M. 1926*a*. Freezing and survival of insects at low temperatures, Quart. Rev. Biol., 1 : 270–82. Pp. 72, 90.

———. 1926*b*. The effect of environmental temperatures upon insect freezing points, Ecology, 7 : 99–106. Pp. 72, 85, 88, 90, 92.

———. 1927*a*. Two factors of heat energy involved in insect cold hardiness, *ibid.*, 8 : 194–96. P. 72.

———. 1927*b*. Measures of insect cold hardiness, Biol. Bull., 52 : 449–57. Pp. 89, 90, 91.

PEARL, R. 1924. Studies in human biology. Baltimore: Williams & Wilkins. P. 8.

———. 1930. Introduction to medical biometry and statistics. Philadelphia: W. B. Saunders Co. P. 46.

PEARL, R., and MINER, J. R. 1935. Experimental studies on the duration of life. XIV. The comparative mortality of certain lower organisms, Quart. Rev. Biol., 10 : 60–75. P. 9.

PEARL, R., and REED, L. J. 1920. On the rate of growth of the population of the United States since 1790 and its mathematical representation, Proc. Nat. Acad. Sc., Washington, 6 : 275–88. P. 44.

PEPPER, J. H., and HASTINGS, E. 1941. Life history and control of the sugar-beet webworm, Bull. Montana Agr. Exper. Sta., No. 389. P. 108.

POLLISTER, A. W., and MOORE, J. A. 1937. Tables for the normal development of *Rana sylvatica,* Anat. Rec., 68 : 489–96. P. 60.

POND, D. D. 1948. Corn earworm, Proc. Pub. Canad. Dept. Agr., Div. Ent., No. 105. P. 93.

POTTS, W. H. 1937. The distribution of tsetse flies in Tanganyika Territory, Bull. Ent. Research, 28 : 129–48. P. 217.

POWSNER, L. 1935. The effects of temperature on the duration of the developmental stages of *Drosophila melanogaster,* Physiol. Zoöl., 8 : 474–520. Pp. 46, 48, 49, 57.

PRESTON, F. W. 1948. The commonness, and rarity, of species, Ecology, 29 : 254–83. P. 238.

RAINEY, R. C., and WALOFF, Z. 1951. Flying locusts and convection currents, Bull. Anti-locust Research Centre London, 9 : 51–70. P. 35.

REIBISCH, J. 1902. Über den Einfluss der Temperatur auf die Entwicklung von Fischeiern, Wissensch. Meeresuntersuch., 2 : 213–31. P. 61.

REYNOLDSON, T. B. 1947a. An ecological study of the enchytraeid worm population of sewage bacteria beds. Field investigations, J. Anim. Ecol., 16 : 26–37. P. 193.

———. 1947b. An ecological study of the enchytraeid worm population of sewage bacteria beds. Laboratory experiments, Ann. Appl. Biol., 34 : 331–45. P. 193.

———. 1948. An ecological study of the enchytraeid worm population of sewage bacteria beds: synthesis of field and laboratory data, J. Anim. Ecol., 17 : 27–38. Pp. 193, 195.

ROBINSON, W. 1928. Water conservation in insects, J. Econ. Ent., 21 : 897–902. Pp. 81, 92.

RYAN, F. J. 1941. Temperature change and the subsequent rate of development, J. Exper. Zoöl., 88 : 25–54. Pp. 53, 54, 55.

SACHAROV, N. L. 1930. Studies on cold resistance in insects, Ecology, 11 : 505–17. Pp. 85, 93, 118.

SALT, G., and HOLLICK, F. S. J. 1946. Studies of wireworm populations. II. Spatial distribution, J. Exper. Biol., 23 : 1–46. Pp. 146, 148.

SALT, R. W. 1936. Studies on the freezing process in insects, Tech. Bull. Minnesota Agr. Exper. Sta., No. 116. Pp. 73, 80, 84, 89, 90, 92.

———. 1950. Time as a factor in the freezing of undercooled insects, *ibid.,* 28 : 285–91. Pp. 84, 86, 87, 93, 94, 115.

SAVAGE, R. M. 1934. The breeding behaviour of the common frog, *Rana temporaria temporaria,* Linn., and of the common toad, *Bufo bufo bufo,* Linn., Proc. Zoöl. Soc. London, 1934, pp. 55–70. P. 113.

———. 1935. The influence of external factors on the spawning date and migration of the common frog, *Rana temporaria temporaria* Linn., *ibid.,* 1935, pp. 49–98. P. 113.

SAVELY, H. E. 1939. Ecological relations of certain animals in dead pine and oak logs, Ecol. Monogr., 9 : 321–85. P. 114.

SCHULMAN, E. 1948. Dendrochronology in north-eastern Utah, Tree-ring Bull., 15 : 2–14. P. 232.

SCOTT, T. G. 1947. Comparative analysis of red fox feeding trends on two central Iowa areas, Research Bull. Iowa Agri. Exper. Sta., No. 353. P. 128.

SEAMANS, H. L. 1923. Forecasting outbreaks of the pale western cutworm in Alberta, Canad. Ent., 55 : 51–53. Pp. 182, 184.

SHELFORD, V. E. 1927. An experimental study of the relations of the codling moth to weather and climate, Bull. Illinois Nat. Hist. Surv., 16 : 311–440. Pp. 58, 62, 66, 88.

SIMPSON, C. B. 1903. The codling moth, Bull. U.S. Div. Ent., No. 41. P. 61.

SIMPSON, G. G. 1952. How many species? Evolution, 6 : 342. P. 252.

SMITH, H. S. 1935. The rôle of biotic factors in the determination of population densities, J. Econ. Ent., 28 : 873–98. Pp. 237, 251.

SOLOMON, M. E. 1949. The natural control of animal populations, J. Anim. Ecol., 18 : 1–35. P. 236.

STEVENSON-HAMILTON, J. 1937. South African Eden: from Sabi game reserve to Kruger national park. London: Cassell & Co., Ltd. P. 252.

STIRRETT, G. M. 1931. Preliminary observations on the winter mortality of the larvae of the European cornborer in Ontario in 1930, 61st Ann. Rep. Ent. Soc. Ontario, pp. 48–52. P. 93.

———. 1938. A study of the flight, oviposition, and establishment periods in the life cycle of the European cornborer *Pyrausta nubilalis* Hbn., and the physical factors affecting them. II. The flight of the European cornborer. Annual cycle of flight. Flight to light trap, Scient. Agr., 18 : 462–84. Pp. 70, 93.

SWEETMAN, H. L. 1929. Precipitation and irrigation as factors in the distribution of the Mexican bean beetle *Epilachna corrupta* Muls., Ecology, 10 : 228–44. P. 91.

———. 1938. Physical ecology of the firebrat, *Thermobia domestica*, Packard, Ecol. Monogr., 8 : 285–311. P. 100.

———. 1939. Responses of the silverfish *Lepisma saccharina*, L. to its physical environment, J. Econ. Ent., 32 : 698–700. P. 100.

TETLEY, J. H. 1937. The distribution of nematodes in the small intestine of the sheep, New Zealand J. Scient. Tech., 18 : 805–17. P. 207.

THOMAS, E. L., and SHEPARD, H. H. 1940. The influence of temperature, moisture, and food on the development and survival of the saw-toothed grain beetle, J. Agr. Research, 60 : 605–15. P. 82.

TRUMBLE, H. C. 1937. The climatic control of agriculture in South Australia, Tr. Roy. Soc. South Australia, 61 : 41–62. P. 165.

TURCĚK, F. J. 1949. The bird populations in some deciduous forests during a gipsy moth outbreak, Bull. Inst. Forestry Research Czechoslovakia, 1949, pp. 108–31. P. 127.

UDVARDY, M. D. F. 1951. The significance of interspecific competition in bird life, Oikos, 3 : 98–123. P. 127.

ULLYETT, G. C. 1947. Mortality factors in populations of *Plutella maculipennis*, Curtiss (Tineidae Lepidoptera) and their relation to the problem of control, Mem. Dept. Agr. South Africa, 2 : 77–202. Pp. 202, 204, 205.

UVAROV, B. P. 1931. Insects and climate, Tr. Ent. Soc. London, 79 : 1–247. Pp. 73, 120.

VANCE, A. M. 1949. Some physiological relations of the female European cornborer moth in controlled environments, J. Econ. Ent., 42 : 474–84. P. 68.

WADLEY, F. M. 1931. Ecology of *Toxoptera graminium*, Ann. Ent. Soc. Amer., 24 : 325–95. Pp. 68, 69, 70.

WATERHOUSE, D. F. 1947. The relative importance of live sheep and of carrion as breeding grounds for the Australian sheep blowfly *Lucilia cuprina*, Bull. Counc. Scient. & Indust. Research Australia, No. 217. P. 115.

WELLINGTON, W. G. 1949a. Temperature measurements in ecological entmology, Nature, 163 : 614–15. Pp. 29, 30.

————. 1949b. The effects of temperature and moisture upon the behaviour of the spruce budworm, *Choristoneura fumiferana* Clemens (Lepidoptera: Tortricidae). I. The relative importance of graded temperatures and rates of evaporation in producing aggregations of larvae, Scient. Agr., 29 : 201–15. Pp. 28, 29.

————. 1949c. The effects of temperature and moisture upon the behaviour of the spruce budworm, *Choristoneura fumiferana* Clemens (Lepidoptera: Torticidae). II. The responses of larvae to gradients of evaporation, *ibid.*, pp. 216–29. P. 29.

————. 1950. Variations in the silk-spinning and locomotor activity of larvae of the spruce budworm *Choristoneura fumiferana* (Clem.), at different rates of evaporation, Tr. Roy. Soc. Canada, 44 : 89–101. Pp. 29, 119.

————. 1952. Air-mass climatology of Ontario north of Lake Huron and Lake Superior before outbreaks of the spruce budworm, *Choristoneura fumiferana* (Clem.), and the forest tent caterpillar, *Malacosoma disstria* Hbn. (Lepidoptera: Tortricidae; Lasiocampidae), Canad. J. Zoöl., 30 : 114–27. Pp. 188, 191.

WELLINGTON, W. G.; FETTES, J. J.; TURNER, K. B.; and BELYEA, R. M. 1950. Physical and biological indicators of the development of outbreaks of the spruce budworm, *Choristoneura fumiferana* (Clem.) (Lepidoptera: Tortricidae), Canad. J. Research, D, 28 : 308–31. Pp. 188, 189, 190, 191.

WIGGLESWORTH, V. B. 1939. The principles of insect physiology. London: Methuen & Co., Ltd. P. 73.

WILKES, A. 1942. The influence of selection on the preferendum of a chalcid (*Microplectron fuscipennis* Zett.) and its significance in the biological control of an insect pest, Proc. Roy. Soc. London, B, 130 : 400–415. P. 28.

WILLIAMS, C. B. 1939. An analysis of four years' captures of insects in the light trap. I. General survey; sex proportion; phenology and time of flight, Tr. Roy. Ent. Soc. London, 89 : 79–132. P. 40.

————. 1940. An analysis of four years' captures of insects in a light trap. II. The effect of weather conditions on insect activity; and the estimation and forecasting of changes in the insect population, *ibid.*, 90 : 227–306. P. 40.

————. 1944. Some applications of the logarithmic series and the index of diversity to ecological problems, J. Ecology, 32 : 1–44. P. 238.

WOODROFFE, G. E., and SOUTHGATE, B. J. 1950. Birds' nests as a source of domestic pests, Proc. Zoöl. Soc. London, 121 : 55–62. P. 115.

ZUMPT, F. 1940. Die Verbreitung der *Glossina palpalis*—Subspezies in Belgisch—Kongogebeit, Rev. Zoöl. & Bot. Africa, 33 : 136–49. P. 217.

Index